Encyclopaedia of
Nanotechnology

Encyclopaedia of
Nanotechnology

Vol. 2 : Synthesis and Characterization

Dr. Rita Khare

AGROTECH PRESS

JAIPUR ▪ NEW DELHI

AGROTECH PRESS

Regd. Off. "Bony Residency", Gate No. 2
Opp. Tilak Public School, Vishweswariya Nagar,
Gopalpura Road, Jaipur - 302018, India
Ph: 0141 - 2761280, Telefax: 2761381
E-mail: agrotechpress@gmail.com

Delhi Office: 102, 1st Floor, Satyam House,
4327/3, Ansari Road, Darya Ganj,
New Delhi - 110002, India
Ph: 011-45652440
visit us: www.agrotechpress.com

Distributed by :
KEYA INTERNATIONAL PRESS LLP
E-mail: keyainternationalpressllp@gmail.com

Published in India, 2024

ISBN : 9789384568566 (10 Vol. Set)

Typeset by Agrotech Press

Printed and bound in India by Replika Press Pvt. Ltd.

Contents

Preface

Nanotechnology is well recognized as a science and technology megatrend for the beginning of the 21st century. Nanotechnology involves the understanding, manipulation, and control of matter at dimensions of roughly 1 to 100 nanometers. Nanotechnology encompasses science, engineering and technology and involves imaging, measuring, modeling, and manipulating matter at the nanoscale. The development of unique nanoscale structures has the potential to revolutionize industry, including electronics, medicine, and consumer products. This encyclopedia aims to make understand where nanotechnology is now - transitioning to complex systems and fundamentally new products - and communicates the societal promise of nanotechnology to specialists and the public. As scientific inquiries in nanoscience progress, the opportunity for nanotechnology breakthroughs continue to expand. The smaller the technology-whether a memory chip for a supercomputer or a tiny machine racing through the body to detect cancer-the greater the chance for a large societal impact. The future of nanotechnology will rely upon the availability of multifunctional nanoplatforms. The development of multi-functional nanosystems not only achieves multi-functionality but also sometimes alleviates the limitations of one of the components. With the different techniques for sysnthesis as described in this book, there will be different results achieved. Care should be taken when the technique is chosen, as the effect on the entire sample will also depend on the composition and the amount of the sample.

The properties of nanoparticles depend on their size, shape, etc., in addition to the composition. Therefore, it is of paramount importance to synthesize nanostructures of greater complexity in order to create newer properties and features within the nano-regime. This enables new applications and further fundamental studies on such systems. Although, a large body of experimental work on the synthesis of nonmaterials has been reported, the synthesis methodologies are mainly trial-and-error in nature due to the poor knowledge about the mechanism of development of nanosystems. The first step towards the generation of complex shape is to prepare particles of controlled shapes. The book explores the exact roles of various capping, protecting and scaffolding agents in the control of particle shapes. We are taking a combined experimental and theoretical approach towards the problem. The top-down approach anticipates nanodevices that must be built piece by piece in stages, much as manufactured items are made. In contrast, bottom-up techniques build or grow larger structures atom by atom or molecule by molecule. The book explores both approaches.

In order to produce faster, cheaper, more efficient, sustainable products, it is necessary to conscientiously study the characteristics of nanomaterials. Materials science today is evolving to the established discipline based on precise measurement and computational modeling. The advanced characterization and measurement technologies responding to the material needs are indispensable for the innovation. This volume aims at is discussing state-of-the-art analytical technologies to clarify the comprehensive characteristics of materials and the development of high resolution and multifunctional characterization technologies, and moreover the in-situ dynamic measurement technologies in the controlled environments related to material creation and functionality expression. Materials at the nanoscale do not always behave as expected and nanomaterials differ from larger materials not just in size, but in surface/interface-to-volume ratio and grain shapes, which are the origins of their unique electrical, optical, thermodynamic, mechanical, and chemical properties. Because of these unique properties, many scientists believe that nanomaterials hold the key to significant future advances in a wide range of miniaturized consumer products such as miniaturized computer chips, nanoscale sensors, and devices for sorting DNA molecules, including products that are reliant on synthesized nanomaterials and their integration with microsystems and biotechnology. To harness the power of materials at the nanoscale, the volume recomends a variety of groundbreaking methods for synthesis and characterization. It includes ion beam and x-ray testing, surface analysis, electronic and near-field microscopy, optical characterization, mechanical properties testing, and sample preparation—a crucial stage in scientific observation at the nanometric scale. As the development and use of nanotechnologies grows, aspects of nanosafety are becoming increasingly important, and the ability to measure and characterize nanoparticles is essential to improved understanding and control. Characterization techniques are therefore high on the agenda of research. The volume deals in detail the materials and biophysical characterization technology and expertise that enables scientists and engineers to understand and control the properties of dispersed systems. This book shows an update in the field of nano fabrications techniques of two and three dimensional structures as well as ultimate three dimensional characterization methods from the atom range to the micro scale. This volume is a comprehensive summary of new trends of nano-fabrication and characterization methods.

Dr. Rita Khare

1

Introduction

Nanomaterials are not simply another step in the miniaturization of materials. They often require very different production approaches. There are several processes to create nanomaterials, classified as 'top-down' and 'bottom-up'. Although many nanomaterials are currently at the laboratory stage of manufacture, a few of them are being commercialized.

Thin Films, Layers and Surfaces

One-dimensional nanomaterials, such as thin films and engineered surfaces, have been developed and used for decades in fields such as electronic device manufacture, chemistry and engineering. In the silicon integrated-circuit industry, for example, many devices rely on thin films for their operation, and control of film thicknesses approaching the atomic level is routine. Monolayers (layers that are one atom or molecule deep) are also routinely made and used in chemistry. The formation and properties of these layers are reasonably well understood from the atomic level upwards, even in quite complex layers (such as lubricants). Advances are being made in the control of the composition and smoothness of surfaces, and the growth of films.

Engineered surfaces with tailored properties such as large surface area or specific reactivity are used routinely in a range of applications such as in fuel cells and catalysts. The large surface area provided by nanoparticles, together with their ability to self assemble on a support surface, could be of use in all of these applications.

Although they represent incremental developments, surfaces with enhanced properties should find applications throughout the chemicals and energy sectors. The benefits could surpass the obvious economic and resource savings achieved by higher activity and greater selectivity in reactors and separation processes, to enabling small-scale distributed processing (making chemicals as close as possible to the point of use). There is already a move in the chemical industry towards this. Another use could be the small-scale, on-site production of high value chemicals such as pharmaceuticals.

Graphene and Other Single- and Few-Layer Materials

Graphene is an atomic-scale honeycomb lattice made of carbon atoms. Graphene is undoubtedly emerging as one of the most promising nanomaterial because of its unique combination of superb properties, which opens a way for its exploitation in a wide spectrum of applications ranging from electronics to optics, sensors, and biodevices.

For instance, graphene-based nanomaterials have many promising applications in energy-related areas. Just some recent examples: Graphene improves both energy capacity and charge rate in rechargeable batteries; activated graphene makes superior supercapacitors for energy storage; graphene electrodes may lead to a promising approach for making solar cells that are inexpensive, lightweight and flexible; and multifunctional graphene mats are promising substrates for catalytic systems

The fascination with atomic-layer materials that has started with graphene has spurred researchers to look for other 2D structures like for instance metal carbides and nitrides. One particularly interesting analogue to graphene would be 2D silicon – silicene – because it could be synthesized and processed using mature semiconductor techniques, and more easily integrated into existing electronics than graphene is currently. Another material of interest is 2D boron, an element with worlds of unexplored potential. And yet another new two-dimensional material – made up of layers of crystal known as molybdenum oxides – has unique properties that encourage the free flow of electrons at ultra-high speeds.

Severe-environment survivability is the key factor in the development of new space materials. These materials must exhibit excellent physical properties accompanied by lightweight, reusability, and multifunctional capabilities, or must be related to processes that involve either low-energy consumption or a highly efficient method of energy storage, conversion or production.

Nanoparticle Synthesis and Characterization

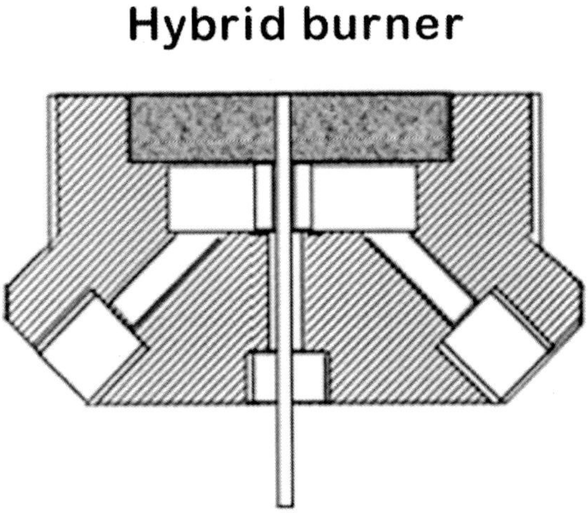

Figure: *Hybrid burner*

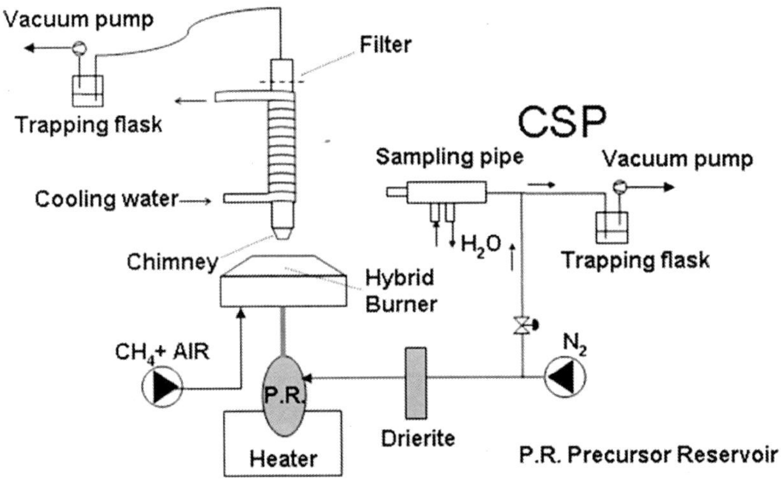

Figure: *The experimental apparatus and two collecting systems*

The interest for the synthesis processes of nanostructured materials, such as nanoparticles, nanotubes and nanolayers, is due to both scientific and technological motivations. The study of nanomaterials allows to investigate totally new phenomena, while their assembly in nanostructures offers the possibility of many applications in numerous technological fields.

Numerous processes are utilised for the synthesis of nanostructured materials and in particular for nanoparticle production. These processes can be divided in three main categories. The mechanical methods obtain a reduction of the particle dimensions through mechanical systems such as the “ball milling” technique. The liquid-chemical methods rely on the precipitation of a solid in a solution or the chemical conversion of colloidal dispersions in a gel ("sol-gel technique"). The high temperature methods for the synthesis are numerous and include evaporation/ condensation techniques, aerosol techniques and flame synthesis. The high temperature synthesis offers several advantages for nanoparticle production in terms of both purity of materials and size control of the particles. For the flame synthesis process there are two main ways to inject the precursor in the flame: in homogenous phase, through an evaporation system at controlled temperature, or through a spray system, in liquid phase. This last method offers a greater flexibility in term of type of precursors to be utilised and flow control determining the amount of produced material. Moreover the easy scaling-up of the experimental equipment allow the production of nanopowders in sufficient quantities to be used for the realisation of prototypes by other research groups.

Nanostructures Obtained by Laser Ablation in Liquids: From Basics to Applications

Laser ablation is the process of removing material from a solid (or occasionally liquid) surface by irradiating it with a laser beam. At low laser flux, the material is heated by the absorbed laser energy and evaporates or sublimates. At high laser flux, the material is typically converted to a plasma. Usually, laser ablation refers to removing material with a pulsed laser, but it is possible to ablate material with a continuous wave laser beam if the laser intensity is high enough.

The depth over which the laser energy is absorbed, and thus the amount of material removed by a single laser pulse, depends on the material's optical properties and the laser wavelength and pulse length. The total mass ablated from the target per laser pulse is usually referred to as ablation rate.

Laser pulses can vary over a very wide range of duration (milliseconds to femtoseconds) and fluxes, and can be precisely controlled. This makes laser ablation very valuable for both research and industrial applications.

Applications

The simplest application of laser ablation is to remove material from a solid surface in a controlled fashion. Laser machining and particularly laser drilling are examples; pulsed lasers can drill extremely small, deep holes through very hard materials. Very short laser pulses remove material so quickly that the surrounding material absorbs very little heat, so laser drilling can be done on delicate or heat-sensitive materials, including tooth enamel (laser dentistry). Several workers have employed laser ablation and gas condensation to produce nano particles of metal, metal oxides and metal carbides.

Also, laser energy can be selectively absorbed by coatings, particularly on metal, so CO_2 or Nd:YAG pulsed lasers can be used to clean surfaces, remove paint or coating, or prepare surfaces for painting without damaging the underlying surface. High power lasers clean a large spot with a single pulse. Lower power lasers use many small pulses which may be scanned across an area. The advantages are:

- No solvents are used, so it is environmentally friendly and operators are not exposed to chemicals.
- It is relatively easy to automate, e.g., by using robots.
- The running costs are lower than dry media or CO_2 ice blasting, although the capital investment costs are much higher.
- The process is gentler than abrasive techniques, e.g. carbon fibres within a composite material are not damaged.
- Heating of the target is minimal.

Another class of applications uses laser ablation to process the material removed into new forms either not possible or difficult to produce by other means. A recent example is the production of carbon nanotubes.

In March 1995 Guo et al. were the first to report the use of a laser to ablate a block of pure graphite, and later graphite mixed with catalytic metal. The catalytic metal can consist of elements such as cobalt, niobium, platinum, nickel, copper, or a binary combination thereof. The composite block is formed by making a paste of graphite powder, carbon cement, and the metal. The paste is next placed in a cylindrical mold and baked for several hours. After solidification, the graphite block is placed inside an oven with a laser pointed at it, and argon gas is pumped along the direction of the laser point. The oven temperature is approximately 1200 °C. As the laser ablates the target, carbon nanotubes form and are carried by the gas flow onto a cool copper collector. Like carbon nanotubes formed using the electric-arc discharge technique, carbon nanotube fibres are deposited in a haphazard and tangled fashion. Single-walled nanotubes are formed from the block of graphite and metal catalyst particles, whereas multi-walled nanotubes form from the pure graphite starting material.

A variation of this type of application is to use laser ablation to create coatings by ablating the coating material from a source and letting it deposit on the surface to be coated; this is a special type of physical vapor deposition called pulsed laser deposition (PLD), and can create coatings from materials that cannot readily be evaporated any other way. This process is used to manufacture some types of high temperature superconductor.

Remote laser spectroscopy uses laser ablation to create a plasma from the surface material; the composition of the surface can be determined by analyzing the wavelengths of light emitted by the plasma.

Propulsion

Finally, laser ablation can be used to transfer momentum to a surface, since the ablated material applies a pulse of high pressure to the surface underneath it as it expands. The effect is similar to hitting the surface with a hammer. This process is used in industry to work-harden metal surfaces, and is one damage mechanism for a laser weapon. It is also the basis of pulsed laser propulsion for spacecraft.

Manufacturing

The laser ablation of electronic semiconductors and microprocessors is now being pioneered in the UK to keep electronic manufacturers designs confidential. The main reason is that it greatly reduces the risk of copying infringements.

Processes are currently being developed to use laser ablation in the removal of thermal barrier coating on high-pressure gas turbine components. Due to the low heat input, TBC removal can be completed with minimal damage to the underlying metallic coatings and parent material.

Applications in Medicine

Laser ablation has biological applications and can be used to destroy nerves and other tissues. For example, a species of pond snail, *Helisoma trivolvis*, can have their sensory neurons laser ablated off when the snail is still an embryo to prevent use of those nerves.

Laser ablation can be used on benign and malignant lesions in various organs, which is called Laser-induced interstitial thermotherapy. The main applications currently involve the reduction of benign thyroid nodules and destruction of primary and secondary malignant liver lesions.Parts Cleaning

Parts cleaning is essential to many industrial processes, as a prelude to surface finishing or to protect sensitive components. Electroplating is particularly sensitive to part cleanliness, since molecular layers of oil can prevent adhesion of the coating. ASTM B322 is a standard guide for cleaning metals prior to electroplating. Cleaning processes include solvent cleaning, hot alkaline detergent cleaning, electrocleaning, and acid etch. The most common industrial test for cleanliness is the waterbreak test, in which the surface is thoroughly rinsed and held vertical. Hydrophobic contaminants such as oils cause the water to bead and break up, allowing the water to drain rapidly. Perfectly clean metal surfaces are hydrophilic and will retain an unbroken sheet of water that does not bead up or drain off. ASTM F22 describes a version of this test. This test does not detect hydrophilic contaminants, but the electroplating process can displace these easily since the solutions are water-based. Surfactants such as soap reduce the sensitivity of the test, so these must be thoroughly rinsed off.

For the activities described here the following terms are often found: metal cleaning, metal surface cleaning, component cleaning, degreasing, parts washing, parts cleaning. These are well established in technical language usage but they have their shortcomings. Metal cleaning can easily be mixed up with refinement of unpurified metals. Metal surface cleaning and metal cleaning do not consider the increasing usage of plastics and composite materials in this sector. The term component cleaning leaves out the cleaning of steel sections and sheets and finally degreasing only describes a part of the topic as in most cases also chips, fines, particles, salts etc. have to be removed.

The terms 'commercial and industrial parts cleaning', 'parts cleaning in craft and industry' or 'commercial parts cleaning' probably best describe this field of activity. There are some specialists who prefer the term 'industrial parts cleaning', because they want to exclude maintenance of buildings, rooms, areas, windows, floors, tanks, machinery, hygiene, hands washing, showers etc.

Elements and Their Interactions

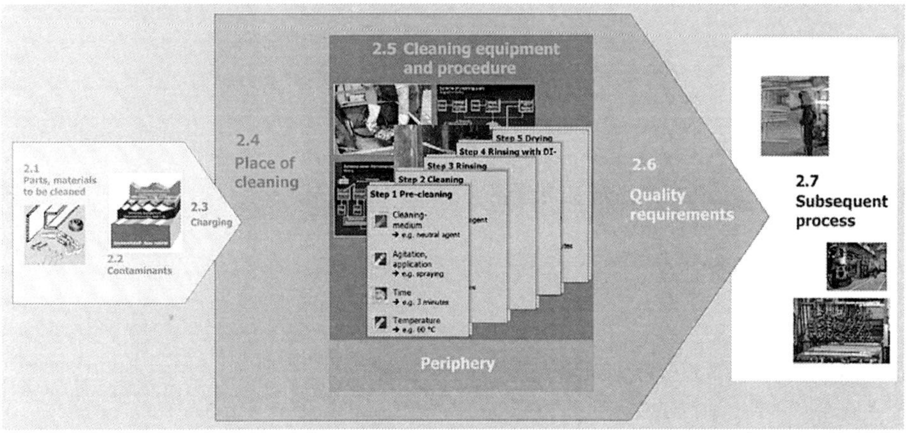

***Figure:** Factors*

Cleaning activities in this sector can only be characterised sufficiently by a description of a number of different factors. These are outlined in illustration 1.

Parts and Materials to be Cleaned

First, consider the parts to be cleaned. They may consist of non- or hardly-processed sections, sheets and wires. But also machined parts or assembled components needing cleaning. Therefore, they may be composed of different metals or different combinations of metals. Plastics and composite materials can frequently be found and indeed are on the increase because e.g. the automobile industry as well as others uses more and more lighter materials.

Mass and size can be very important for the selection of cleaning methods, for example big shafts for ships are usually cleaned manually, whereas tiny shafts for electrical appliances are often cleaned in bulk in highly automated plants.

Similarly important is the geometry of the parts. Long, thin, branching, threaded holes, which could contain jammed chips, feature among the greatest challenges in this technical field. High pressure and the power wash process are one way to remove these chips as well as robots, which are programmed to exactly flush the drilled holes under high pressure.

Contaminations

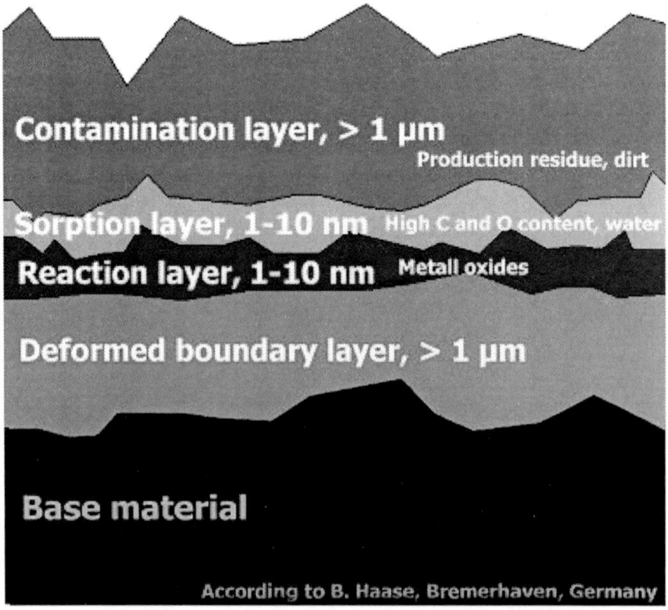

Figure: *Structure of a metallic surface*

The parts are usually covered by unwanted substances, the contaminations or soiling. The definition used is quite varied. In certain cases these coverings may be desired: e.g. one may not wish to remove a paint layer but only the material on top. In another case, where crack proofing is necessary one has to remove the paint layer and it is regarded as an unwanted substance.

The classification of soiling follows the layer structure starting from the base material:

- Deformed boundary layer, > 1 μm
- Reaction layer, 1 - 10 nm
- Sorption layer, 1 - 10 nm
- Contamination layer, > 1 μm

The nearer the layers are to the substrate surface, the more energy is needed to remove them. Correspondingly the cleaning itself can be structured according to the type of energy input:

- mechanical - abrasive: blasting, grinding
- mechanical - non-abrasive: stirring, mixing, ultrasound, spraying
- thermal - reactive: heat treatment much above 100°C in reactive gases
- thermal - non-reactive: temperature below 100°C, increased bath temperature, vapour degreasing
- chemical - abrasive/reactive: pickling in liquids, plasma-assisted, sputter-cleaning, elektropolishing
- chemical - non-reactive: organic solvents, aqueous solutions, supercritical CO_2

The contamination layer may then be further classified according to:

- Origin
- Composition: e.g. cooling lubricants may be composed very differently, thereby single components may account for big problems especially for job shop cleaners, who have no control over prior processes and thus don't know the contaminants. For example silicates may obstruct nitriding.
- State of aggregation
- Chemical and physical properties

The American Society for Testing and Materials (ASTM) presents six groups of contaminations in their manual "Choosing a cleaning process" and relates them to the most common cleaning methods, thereby the suitability of cleaning methods for the removal of a given contaminate is discussed in detail.

In addition they list exemplary cleaning processes for different typical applications. Since one has to consider very many different aspects when choosing a process, this can only serve as a first orientation.

The groups of contaminants are stated as follows:

- Pigmented drawing compounds
- Unpigmented oil and grease
- Chips and cutting fluids
- Polishing and buffing compounds
- Rust and scale
- Others

Charging

In order to select suitable equipment and media it should be known also which amount and which throughput has to be handled. Small amounts can hardly be cleaned economically in larger plants. Also the type of charging has to be ascertained.

Sensitive parts sometimes need to be fixed in boxes. Very economically when dealing with large amounts is bulk charging, but it is quite difficult to achieve a sufficient level of cleanliness with flat pieces clinging together. Also drying can be difficult in these cases.

Place of Cleaning

Another consideration is the place of cleaning. E.g. is the cleaning to be done on site, which can be the case with repair and maintenance work. Usually the cleaning takes place in a workshop.

Several common methods are solvent degreasing, vapor degreasing and using an aqueous parts washer. Companies often want the charging, loading and unloading to be integrated into the production line, which is much more demanding as regards size and throughput ability of the cleaning system.

Such cleaning systems often exactly match the requirements regarding parts, contaminants and charging methods (special production). Nonetheless central cleaning equipment, often built as multi task systems, are commonly used. These systems can suit different cleaning requirements.

Typical examples are the wash stands or the small cleaning machines which are found in many industrial plants.

Cleaning Equipment and Procedure

First, one can differentiate between the following techniques:

- Manual
- Mechanical
- Automatic
- Robot supported

The process may be performed in one step, which is especially true for the manual cleaning, but typically it requires several steps. Therefore, it is not uncommon to find 10 to 20 steps in large plants e.g. for the medical and optical industry. This can be especially complex because non-cleaning steps may be integrated in such plants like application of corrosion protection layers or phosphating. Cleaning can also be simple, the cleaning processes are integrated into other processes as it is the case with electroplating or galvanising, where it usually serves as a pre-treatment step.

The following procedure is quite common:

1. Pre cleaning
2. Main cleaning
3. Rinsing
4. Rinsing with deionised water
5. Rinsing with corrosion protection
6. Drying

Each of these steps may take place in its own bath or chamber or in case of spray cleaning in its own zone (line or multi-chamber equipment). But quite often these steps may have a single chamber into which the respective media are pumped in (single chamber plant).

Besides equipment and procedure, cleaning media plays an important role as it removes the contaminants from the substrate.

For liquid media the following cleaners are in use: aqueous agents, semi-aqueous agents (an emulsion of solvents and water), hydrocarbon based solvents and halogenated solvents. Usually the latter are referred to as chlorinated agents, but there are also brominated and fluorated substances in (limited) use, that is why we have chosen the higher level classification. The hazardous traditionally used chlorinated agents TCE and PCE are nowadays only applied in airtight plants and the modern volume shift systems limits any emissions. In the group of hydrocarbon based solvents, there are some newly developed agents like fatty acid esters made of natural fats and oils, modified alcohols and dibasic esters. Aqueous cleaners are mostly a combination of various substances like alkaline builders, surfactants, sequestering agents, etc. In the case of ferrous metal cleaning, rust inibitors are built into the aqueous cleaner to prevent flash rusting after washing. Their use is on the rise as their results have proven to be in many cases as good or better than hydrocarbon cleaners. Additionally, the wastes generated are less hazardous resulting in less costly disposal.

Aqueous cleaners have advantages as regards to particle and polar contaminants and only require higher inputs of mechanical and thermal energy to be effective, whereas solvents more easily remove oils and greases but have health and environmental risks. In addition most solvents are flammable and create fire and explosion hazards. Nowadays, with the proper industrial parts washer equipment, it is generally accepted that aqueous cleaners remove oil and grease as easily as solvents.

Another approach is with solid cleaning media (blasting) which consists of the CO_2 dry ice process: For tougher requirements pellets are used while for more sensitive materials or components CO_2 in form of snow is applied. One draw back is the high energy consumption required to make dry ice.

Last but not least there are processes without any media like vibration, laser, brushing and blow/exhaust systems.

All cleaning steps are characterised by media and applied temperatures and their individual agitation/application (mechanical impact). There is a wide range of different methods and combinations of these methods:

- Sprinkling
- Spraying
- Power Wash Process
- Blasting
- Flooding
- Movement of parts (turning, oscillating, pivoting)
- Circulation of bath
- Gas or air injection into bath
- Boiling under pressure
- Injection flooding
- Pressure flooding
- Hydroson

Ultrasonic Cleaning

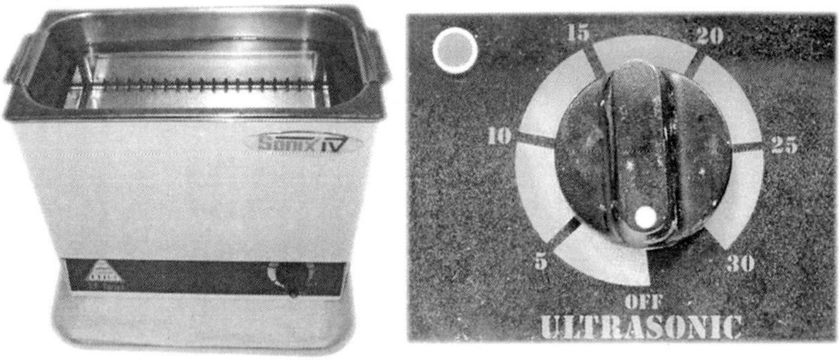

Figure: *Ultrasonic cleaner showing the removable basket in place, and a closeup of the light and timer*

An ultrasonic cleaner is a cleaning device that uses ultrasound (usually from 20–400 kHz) and an appropriate cleaning solvent (sometimes ordinary tap water) to clean delicate items. The ultrasound can be used with just water, but use of a solvent appropriate for the item to be cleaned and the soiling enhances the effect. Cleaning normally lasts between three and six minutes, but can also exceed 20 minutes, depending on the object to be cleaned.

Ultrasonic cleaners are used to clean many different types of objects, including jewellery, lenses and other optical parts, watches, dental and surgical instruments, tools, coins, fountain pens, golf clubs, window blinds, firearms, musical instruments, industrial parts and electronic equipment. They are used in many jewellery workshops, watchmakers' establishments, and electronic repair workshops.

Ultrasonic cleaning uses Cavitation bubbles induced by high frequency pressure (sound) waves to agitate a liquid. The agitation produces high forces on contaminants adhering to substrates like metals, plastics, glass, rubber, and ceramics. This action also penetrates blind holes, cracks, and recesses. The intention is to thoroughly remove all traces of contamination tightly adhering or embedded onto solid surfaces. Water or other solvents can be used, depending on the type of contamination and the workpiece. Contaminants can include dust, dirt, oil, pigments, rust, grease, algae, fungus, bacteria, lime scale, polishing compounds, flux agents, fingerprints, soot wax and mold release agents, biological soil like blood, and so on. Ultrasonic cleaning can be used for a wide range of workpiece shapes, sizes and materials, and may not require the part to be disassembled prior to cleaning. Objects must not be allowed to rest on the bottom of the device during the cleaning process, because that will prevent cavitation from taking place on the part of the object not in contact with water.

In an ultrasonic cleaner, the object to be cleaned is placed in a chamber containing a suitable solution (in an aqueous or organic solution, depending on the application). In aqueous cleaners, the chemical added is a surfactant (eg- laundry detergent) which breaks down the surface tension of the water base. An ultrasound generating transducer built into the chamber, or lowered into the fluid, produces ultrasonic waves in the fluid by changing size in concert with an electrical signal oscillating at ultrasonic frequency. This creates compression waves in the liquid of the tank which 'tear' the liquid apart, leaving behind many millions of microscopic 'voids' or 'partial vacuum bubbles' (cavitation). These bubbles collapse with enormous energy; temperatures and pressures on the order of 5,000 K and 20,000 lbs per square inch are achieved; however, they are so small that they do no more than clean and remove surface dirt and contaminants. The higher the frequency, the smaller the nodes between the cavitation points, which allows for cleaning of more intricate detail.

Figure: *Ultrasonic transducers showing ~20 kHz and ~40 kHz stacks. The active elements (near the top) are two rings of lead zirconate titanate, which are bolted to an aluminium coupling horn.*

Transducers are usually piezoelectric (e.g. made with lead zirconate titanate (PZT), barium titanate, etc), but are sometimes magnetostrictive. The often harsh chemicals used as cleaners in many industries are not needed, or used in much lower concentrations, with ultrasonic agitation. Ultrasonics are used for industrial cleaning, and also used in many medical and dental techniques and industrial processes.

Most hard, non-absorbent materials (metals, plastics, etc.) not chemically attacked by the cleaning fluid are suitable for ultrasonic cleaning. Ideal candidates for ultrasonic cleaning include small electronic parts, cables, rods, wires and detailed items, as well as objects made of glass, plastic, aluminum or ceramic.

Ultrasonic cleaning does not sterilize the objects being cleaned, because spores and viruses will remain on the objects after cleaning.

In medical applications, sterilization normally follows ultrasonic cleaning as a separate step.

Industrial ultrasonic cleaners are used in the automotive, sporting, printing, marine, medical, pharmaceutical, electroplating, disk drive components, engineering and weapons industries.

Megasonic

Megasonic Cleaning is a type of acoustic cleaning, related to ultrasonic cleaning. It is a gentler cleaning mechanism, less likely to cause damage, and is used in wafer, medical implant, and industrial part cleaning. Similar to ultrasonic cleaning, megasonics utilises a transducer that usually sits atop a piezoelectric substrate. The transducer creates an acoustic field at a much higher frequency (typically 0.8–2 MHz) compared to ultrasonic cleaning (20-200 kHz). As a result, the cavitation that occurs is gentler and on a much smaller scale. Megasonics are currently used mainly in the silicon industry.

Megasonics cleaning compared to ultrasonic cleaning - The difference between ultrasonic cleaning and megasonics cleaning lies in the frequency that is used to generate the acoustic waves. Ultrasonic cleaning uses lower frequencies; it produces random cavitation. Megasonics cleaning uses higher frequencies at 1000 kHz; it produces controlled cavitation.

An important distinction between the two methods is that the higher megasonic frequencies do not cause the violent cavitation effects found with ultrasonic frequencies. This significantly reduces or eliminates cavitation erosion and the likelihood of surface damage to the product being cleaned. Parts that would be damaged by ultrasonic frequencies or cavitation effects can often be cleaned without damage in a megasonic bath using the same solution.

With ultrasonics, cavitation occurs throughout the tank, and all sides of submerged parts are cleaned. With megasonics, only the side of the part that is facing the transducer(s) is cleaned. Finally, every cleaning step is described by the time which the parts to be cleaned spends in the respective zone, bath or chamber and thus medium, temperature and agitation can impact on the contamination.

Every cleaning equipment needs a so-called periphery. This term describes measures and equipment on the one hand side to maintain and control baths and on the other hand side to protect human beings and the environment.

In most plants the cleaning agents are circulated until their cleaning power has eventually decreased and reached the maximum tolerable contaminant level. In order to delay the necessary bath exchange as much as possible there are sophisticated treatment attachments in use, removing contaminants and the used up agents from the system. At the same time fresh cleaning agents or parts thereof have to be supplemented, which requires a bath control. The latter is more and more facilitated online and thus allows a computer aided adjustment of the bath. With the help of oil separators, demulsifying agents and evaporators aqueous processes can be conducted 'waste water free'. Complete exchange of baths becomes only necessary every 3 to 12 months.

When using organic solvents the preferred method to achieve a long operating bath life is distillation, an especially effective method to separate contaminants and agents.

The periphery also includes measures to protect the workers like encapsulation, automatic shut off of power supply, automatic refill and sharpening of media (e.g. gas shuttle technique), explosion prevention measures, exhaust ventilation etc., and also measures to protect the environment, e.g. capturing of volatile solvents, impounding basins, extraction, treatment and disposal of resulting wastes. Solvents based cleaning processes have the advantage that the dirt and the cleaning agent can be more easily separated, whereas in aqueous processes is more complex.

In processes without cleaning media like laser ablation and vibration cleaning, only the removed dirt has to be disposed of as there is no cleaning agent. Quite little waste is generated in processes like CO_2 blasting and automatic brush cleaning at the expense of higher energy costs.

Quality Requirements

A standardisation of the quality requirements for cleaned surfaces regarding the following process (e.g. coating, heat treatment) or from the point of view of technical functionality is difficult. However it is possible to use general classifications. In Germany it was attempted to define cleaning as a sub category of metal treatment (DIN 8592: Cleaning as sub category of cutting processes), but this does not cope with all the complexities of cleaning.

The rather general rules includes the classification in intermediate cleaning, final cleaning, precision cleaning and critical cleaning (s. table), in practice seen only as a general guideline.

(1) Related to the total dirt; (2) Only related to Carbon

Thus in practice the rule of thumb is still followed, stating that the quality requirements are met, if the subsequent process does not cause any problems, for example a paint coating does not flake off before the guarantee period ends.

Terms	***Max. allowed dirt***	***Soils removed***	***Explanations***
Intermediate cleaning			E.g. in metal cutting manufacturing
Final cleaning	≤ 500 mg / m² (1)	Mil-sized particles and residues thicker than a monolayer	E.g. before assembling or coating
• Parts for phosphating, painting, enamelling	• 500 – ≤ 5 mg C / m² (2)		
• Parts for case-hardening, nitriding, nitro carburising resp. vacuum treatment	• 500 – ≤ 5 mg C / m² (2)		
• Parts for electroplating, electronic parts	• 20 – ≤ 5 mg C / m² (2)		
Precision cleaning	≤ 50 mg / m² (1)	Supermicrometre particles and residues *thinner* than a monolayer	Controlled environment (Durkee)
Critical cleaning	≤ 5 mg / m² (1)	Sub-micrometre particles and non-volatile residue measured in Angstroms	cleanroom (Durkee)

Where this is not sufficient, especially in case of external orders, because of missing standards there are often specific customer requirements regarding remaining contamination, corrosion protection, spots and gloss level etc.

Measuring methods to ensure quality therefore do not play a bigger role in the workshops, although there exist a broad scale of different methods, from visual control over simple testing methods (among other things water break test, wipe test, measurement of contact angle, test inks, tape test) to complex analysis methods (among others gravimetric test, particle counting, infrared spectroscopy, glow discharge spectroscopy, energy dispersive X-ray analysis, scanning electron microscopy and electrochemical methods). Nevertheless there are only few methods, which can be applied directly in the line and which offer reproducible and comparable results. It was not until recently that bigger advancements in this area have been made

The general situation has changed meanwhile, because of dramatically rising cleanliness requirements for certain components in the automotive industry. For example brake systems and fuel-injection systems need to be fitted with increasingly smaller diameters and they have to withstand increasingly higher pressures. Therefore, a very minor particle contamination may lead to big problems. Due to the rising innovation speed, the industry cannot afford to identify possible failures at a relatively late stage. Therefore, the standard VDA 19/ISO 16232 'Road Vehicles – Cleanliness of Components of Fluid Circuits' was developed which describes methods that can control the compliance with the cleanliness requirements.

Subsequent Process

When choosing cleaning techniques, cleaning agents and cleaning processes, the subsequent processes, i.e. the further processing of the cleaned parts is of special interest.

The classification follows basically the metal work theory:

- machining
- cutting
- joining
- coating
- heat treatment
- assembling
- measuring, testing
- repairing, maintenance.

In the course of time empirical values were established, how efficient the cleaning has to be, to assure the processes for the particular guarantee period and beyond. Choosing the cleaning method often starts from here.

Challenges and Trends

The details above illustrate how extremely complex this specific field is. Already small changes in the requirements can necessitate completely different processes. Thus it defies scientific technical determination. On the other hand it becomes more and more important to receive the required degree of cleanliness as cost-effective as possible and with continuously minimised health and environmental risks, because cleaning has become of central importance for the supply chain in manufacturing. Applying companies usually rely on their suppliers, who—due to a big experience base—suggest adequate equipment and processes, which are then adapted to the detailed requirements in tests stations at the supplier's premises. However they are limited to their scope of technology.

Matrix-Assisted Laser Desorption/Ionization

Matrix-assisted laser desorption/ionization (MALDI) is a soft ionization technique used in mass spectrometry, allowing the analysis of biomolecules (biopolymers such as DNA, proteins, peptides and sugars) and large organic molecules (such as polymers, dendrimers and other macromolecules), which tend to be fragile and fragment when ionized by more conventional ionization methods.

It is similar in character to electrospray ionization (ESI) in that both techniques are relatively soft ways of obtaining large ions in the gas phase, though MALDI produces far fewer multiply charged ions.

MALDI is a two-step process. First, desorption is triggered by a UV laser beam. Matrix material heavily absorbs UV laser light, leading to the ablation of upper layer (~1 μm) of the matrix material.

The hot plume produced during ablation contains many species: neutral and ionized matrix molecules, protonated and deprotonated matrix molecules, matrix clusters and nanodroplets. Second, the analyte molecules are ionized (more accurately protonated or deprotonated) in the hot plume. Ablated species may participate in the ionization of analyte, though the mechanism of MALDI is still debated.

Matrix

UV MALDI Matrix List				
Compound	***Other Names***	***Solvent***	***Wavelength (nm)***	***Applications***
2,5-dihydroxy benzoic acid	DHB, Gentisic acid	acetonitrile, water, methanol, acetone, chloroform	337, 355, 266	peptides, nucleotides, oligonucleotides, oligosaccharides
3,5-dimethoxy-4-hydroxycinnamic acid	sinapic acid; sinapinic acid; SA	acetonitrile, water, acetone, chloroform	337, 355, 266	peptides, proteins, lipids
4-hydroxy-3-methoxycinnamic acid[2][3]	ferulic acid	acetonitrile, water, propanol	337, 355, 266	proteins
α-Cyano-4-hydroxycinnamic acid	CHCA	acetonitrile, water, ethanol, acetone	337, 355	peptides, lipids, nucleotides
Picolinic acid	PA	Ethanol	266	oligonucleotides
3-hydroxy picolinic acid	HPA	Ethanol	337, 355	oligonucleotides

The matrix consists of crystallized molecules, of which the three most commonly used are 3,5-dimethoxy-4-hydroxycinnamic acid (sinapinic acid), α-cyano-4-hydroxycinnamic acid (CHCA, alpha-cyano or alpha-matrix) and 2,5-dihydroxybenzoic acid (DHB). A solution of one of these molecules is made, often in a mixture of highly purified water and an organic solvent (normally acetonitrile (ACN) or ethanol). Trifluoroacetic acid (TFA) may also be added. A good example of a matrix-solution would be 20 mg/mL sinapinic acid in ACN:water:TFA (50:50:0.1).

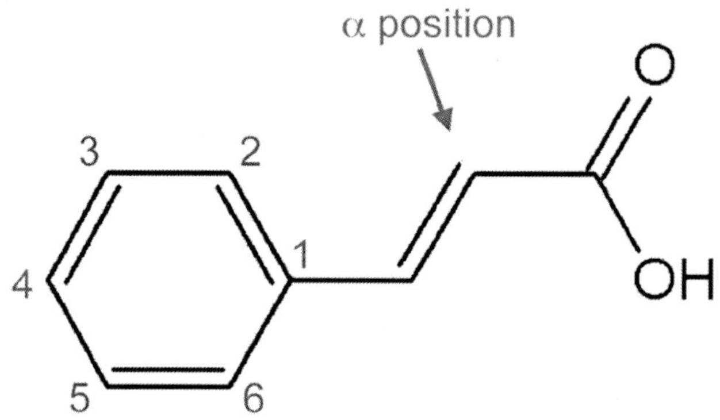

Figure: *Notation for cinnamic acid substitutions.*

The identification of suitable matrix compounds is determined to some extent by trial and error, but they are based on some specific molecular design considerations:

- They are of a fairly low molecular weight (to allow easy vaporization), but are large enough (with a low enough vapor pressure) not to evaporate during sample preparation or while standing in the spectrometre.
- They are often acidic, therefore act as a proton source to encourage ionization of the analyte. Basic matrices have also been reported.
- They have a strong optical absorption in either the UV or IR range, so that they rapidly and efficiently absorb the laser irradiation. This efficiency is commonly associated with chemical structures incorporating several conjugated double bonds, as seen in the structure of cinnamic acid.
- They are functionalized with polar groups, allowing their use in aqueous solutions.
- They typically contain a chromophore.

The matrix solution is mixed with the analyte (e.g. protein-sample). A mixture of water and organic solvent allows both hydrophobic and water-soluble (hydrophilic) molecules to dissolve into the solution. This solution is spotted onto a MALDI plate (usually a metal plate designed for this purpose). The solvents vaporize, leaving only the recrystallized matrix, but now with analyte molecules embedded into MALDI crystals. The matrix and the analyte are said to be co-crystallized. Co-crystallization is a key issue in selecting a proper matrix to obtain a good quality mass spectrum of the analyte of interest.

Laser

MALDI techniques typically employ the use of UV lasers such as nitrogen lasers (337 nm) and frequency-tripled and quadrupled Nd:YAG lasers (355 nm and 266 nm respectively). Although not as common, infrared lasers are used due to their softer mode of ionization. IR-MALDI also has the advantage of greater material removal (useful for biological samples), less low-mass interferences, and compatibility with other matrix-free laser desorption mass spectrometry methods.

Ionization Mechanism

The laser is fired at the matrix crystals in the dried-droplet spot. The matrix absorbs the laser energy and it is thought that primarily the matrix is desorbed and ionized (by addition of a proton) by this event. The matrix is then thought to transfer proton to the analyte molecules (e.g., protein molecules), thus charging the analyte. Ions observed after this process consist of a neutral molecule [M] and an added or removed ion. Together, they form a quasimolecular ion, for example $[M+H]^+$ in the case of an added proton, $[M+Na]^+$ in the case of an added sodium ion, or $[M-H]^-$ in the case of a removed proton. MALDI is capable of creating singly charged ions, but multiply charged ions ($[M+nH]^{n+}$) can also be created, as a function of the matrix, the laser intensity and/or the voltage used. Note that these are all even-electron species. Ion signals of radical cations (photoionized molecules) can be observed, e.g., in case of matrix molecules and other organic molecules.

Atmospheric Pressure Matrix-Assisted Laser Desorption/Ionization

Atmospheric pressure (AP) matrix-assisted laser desorption/ ionization (MALDI) is an ionization technique (ion source) that in contrast to vacuum MALDI operates at normal atmospheric environment. The main difference between vacuum MALDI and AP-MALDI is the pressure in which the ions are created. In vacuum MALDI, ions are typically produced at 10 mTorr or less while in AP-MALDI ions are formed in atmospheric pressure. In the past the main disadvantage of AP MALDI technique compared to the conventional vacuum MALDI has been its limited sensitivity; however, ions can be transferred into the mass spectrometre with high efficiency and attomole detection limits have been reported.

AP-MALDI is used in mass spectrometry (MS) in a variety of applications ranging from proteomics to drug discovery. Popular topics that are addressed by AP-MALDI mass spectrometry include: proteomics; mass analysis of DNA, RNA, PNA, lipids, oligosaccharides, phosphopeptides, bacteria, small molecules and synthetic polymers, similar applications as available also for vacuum MALDI instruments.

The AP-MALDI ion source is easily coupled to an ion trap mass spectrometre or any other MS system equipped with ESI (electrospray ionization) or nanoESI source.

Mass Spectrometre

Figure: *Sample target for a MALDI mass spectrometre*

The type of a mass spectrometre most widely used with MALDI is the TOF (time-of-flight mass spectrometre), mainly due to its large mass range.

The TOF measurement procedure is also ideally suited to the MALDI ionization process since the pulsed laser takes individual 'shots' rather than working in continuous operation. MALDI-TOF instrument or reflectron is equipped with an "ion mirror" that reflects ions using an electric field, thereby doubling the ion flight path and increasing the resolution.

Today, commercial reflectron TOF instruments reach a resolving power m/Δm of well above 20,000 FWHM (full-width half-maximum, Δm defined as the peak width at 50% of peak height).

MALDI has been coupled with IMS-TOF MS to identify phosphorylated and non-phosphorylated peptides.

MALDI-FT-ICR MS has been demonstrated to be a useful technique where high resolution MALDI-MS measurements are desired.

History

The term matrix-assisted laser desorption ionization (MALDI) was coined in 1985 by Franz Hillenkamp, Michael Karas and their colleagues. These researchers found that the amino acid alanine could be ionized more easily if it was mixed with the amino acid tryptophan and irradiated with a pulsed 266 nm laser. The tryptophan was absorbing the laser energy and helping to ionize the non-absorbing alanine. Peptides up to the 2843 Da peptide melittin could be ionized when mixed with this kind of "matrix". The breakthrough for large molecule laser desorption ionization came in 1987 when Koichi Tanaka of Shimadzu Corp. and his co-workers used what they called the "ultra fine metal plus liquid matrix method" that combined 30 nm cobalt particles in glycerol with a 337 nm nitrogen laser for ionization.

Using this laser and matrix combination, Tanaka was able to ionize biomolecules as large as the 34,472 Da protein carboxypeptidase-A. Tanaka received one-quarter of the 2002 Nobel Prize in Chemistry for demonstrating that, with the proper combination of laser wavelength and matrix, a protein can be ionized. Karas and Hillenkamp were subsequently able to ionize the 67 kDa protein albumin using a nicotinic acid matrix and a 266 nm laser. Further improvements were realised through the use of a 355 nm laser and the cinnamic acid derivatives ferulic acid, caffeic acid and sinapinic acid as the matrix. The availability of small and relatively inexpensive nitrogen lasers operating at 337 nm wavelength and the first commercial instruments introduced in the early 1990s brought MALDI to an increasing number of researchers. Today, mostly organic matrices are used for MALDI mass spectrometry.

Applications

Biochemistry

In proteomics, MALDI is used for the rapid identification of proteins isolated by using gel electrophoresis: SDS-PAGE, size exclusion chromatography, affinity chromatography, strong/weak ion exchange, isotope coded protein labelling (ICPL), and two-dimensional gel electrophoresis. Peptide mass fingerprinting is the most popular analytical application of MALDI-TOF mass spectrometres. MALDI TOF/TOF mass spectrometres are used to reveal amino acid sequence of peptides using post-source decay or high energy collision-induced dissociation.

Loss of sialic acid has been identified in papers when DHB has been used as a matrix for MALDI MS analysis of glycosylated peptides. Using sinapinic acid, 4-HCCA and DHB as matrices, S. Martin studied loss of sialic acid in glycosylated peptides by metastable decay in MALDI/TOF in linear mode and reflector mode. A group at SHIMADZU CORPORATION derivatized the sialic acid by an amidation reaction as a way to improve detection sensitivity and also demonstrated that ionic liquid matrix reduces a loss of sialic acid during MALDI/TOF MS analysis of sialylated oligosaccharides. THAP, DHAP, and a mixture of 2-aza-2-thiothymine and phenylhydrazine have been identified as matrices that could be used to minimize loss of sialic acid during MALDI MS analysis of glycosylated peptides. It has been reported that a reduction in loss of some post-translational modifications can be accomplished if IR MALDI is used instead of UV MALDI.

In molecular biology, a mixture of 5-methoxysalicylic acid and spermine can be used as a matrix for oligonucleotides analysis in MALDI mass spectrometry, for instance after oligonucleotide synthesis.

Organic Chemistry

Some synthetic macromolecules, such as catenanes and rotaxanes, dendrimers and hyperbranched polymers, and other assemblies, have molecular weights extending into the thousands or tens of thousands, where most ionization techniques have difficulty producing molecular ions. MALDI is a simple and fast analytical method that can allow chemists to rapidly analyze the results of such syntheses and verify their results.

Polymer Chemistry

In polymer chemistry MALDI can be used to determine the molar mass distribution. Polymers with polydispersity greater than 1.2 are difficult to characterize with MALDI due to the signal intensity discrimination against higher mass oligomers. A good matrix for polymers is dithranol and AgTFA. The sample must first be mixed with dithranol and the AgTFA added afterwards; otherwise the sample would precipitate out of solution.

Microbiology

MALDI/TOF spectra are used for the identification of microorganisms such as bacteria or fungi. A colony of the microbe in question is smeared directly on the sample target and overlaid with matrix. The mass spectra generated are analyzed by dedicated software and compared with stored profiles. Species diagnosis by this procedure is much faster, more accurate and cheaper than other procedures based on immunological or biochemical tests. MALDI/TOF may become the standard method for species identification in medical microbiological laboratories over the next few years.

Reproducibility and Performance

The sample preparation for MALDI is important for both sensitivity, reproducibility, and quantification of mass analysis. Inorganic salts which are also part of protein extracts interfere with the ionization process. The salts can be removed by solid phase extraction or by washing the dried-droplet MALDI spots with cold water. Both methods can also remove other substances from the sample. The matrix-protein mixture is not homogenous because the polarity difference leads to a separation of the two substances during co-crystallization. The spot diameter of the target is much larger than that of the laser, which makes it necessary to make many laser shots at different places of the target, to get the statistical average of the substance concentration within the target spot. The matrix chemical composition, the addition of trifluoroacetic acid, formic acid, fructose, delay time between the end of laser pulse and start of ion acceleration in the ion source (in vacuum MALDI sources), laser wavelength, UV energy (as well as its density and homogeneity) in a focused light spot produced by pulsed laser, and the impact angle of the laser on the target are among critical parameters for the quality and reproducibility of the MALDI-TOF MS method.

Laser Induced Breakdown Spectroscopy

Laser-induced breakdown spectroscopy (LIBS) is a type of atomic emission spectroscopy which uses a highly energetic laser pulse as the excitation source. The laser is focused to form a plasma, which atomizes and excites samples. In principle, LIBS can analyse any matter regardless of its physical state, be it solid, liquid or gas. Because all elements emit light of characteristic frequencies when excited to sufficiently high temperatures, LIBS can (in principle) detect all elements, limited only by the power of the laser as well as the sensitivity and wavelength range of the spectrograph & detector. If the constituents of a material to be analyzed are known, LIBS may be used to evaluate the relative abundance of each constituent element, or to monitor the presence of impurities. In practice, detection limits are a function of a) the plasma excitation temperature, b) the light collection window, and c) the line strength of the viewed transition. LIBS makes use of optical emission spectrometry and is to this extent very similar to arc/spark emission spectroscopy.

LIBS operates by focusing the laser onto a small area at the surface of the specimen; when the laser is discharged it ablates a very small amount of material, in the range of nanograms to picograms, which generates a plasma plume with temperatures in excess of 100,000 K. During data collection, typically after local thermodynamic equilibrium is established, plasma temperatures range from 5,000–20,000 K. At the high temperatures during the early plasma, the ablated material dissociates (breaks down) into excited ionic and atomic species. During this time, the plasma emits a continuum of radiation which does not contain any useful information about the species present, but within a very small timeframe the plasma expands at supersonic velocities and cools. At this point the characteristic atomic emission lines of the elements can be observed. The delay between the emission of continuum radiation and characteristic radiation is in the order of 10 μs, this is why it is necessary to temporally gate the detector.

LIBS can often be referred to as its alternative name: laser-induced plasma spectroscopy (LIPS). The term LIPS has alternative meanings that are outside the field of analytical spectroscopy, therefore the term LIBS is preferred.

LIBS is technically very similar to a number of other laser-based analytical techniques, sharing much of the same hardware. These techniques are the vibrational spectroscopic technique of Raman spectroscopy, and the fluorescence spectroscopic technique of laser-induced fluorescence (LIF). In fact devices are now being manufactured which combine these techniques in a single instrument, allowing the atomic, molecular and structural characterisation of a specimen as well as giving a deeper insight into physical properties.

Design

A typical LIBS system consists of a Nd:YAG solid-state laser and a spectrometre with a wide spectral range and a high sensitivity, fast response rate, time gated detector. This is coupled to a computer which can rapidly process and interpret the acquired data. As such LIBS is one of the most experimentally simple spectroscopic analytical techniques, making it one of the cheapest to purchase and to operate.

The Nd:YAG laser generates energy in the near infrared region of the electromagnetic spectrum, with a wavelength of 1064 nm. The pulse duration is in the region of 10 ns generating a power density which can exceed 1 $GW \cdot cm^{-2}$ at the focal point. Other lasers have been used for LIBS, mainly the Excimer (Excited dimer) type that generates energy in the visible and ultraviolet regions.

The spectrometre consists of either a monochromator (scanning) or a polychromator (non-scanning) and a photomultiplier or CCD detector respectively. The most common monochromator is the Czerny-Turner type whilst the most common polychromator is the Echelle type. However, even the Czerny-Turner type can be (and is often) used to disperse the radiation onto a CCD effectively making it a polychromator. The polychromator spectrometre is the type most commonly used in LIBS as it allows simultaneous acquisition of the entire wavelength range of interest.

The spectrometre collects electromagnetic radiation over the widest wavelength range possible, maximising the number of emission lines detected for each particular element. Spectrometre response is typically from 1100 nm (near infrared) to 170 nm (deep ultraviolet), the approximate response range of a CCD detector. All elements have emission lines within this wavelength range. The energy resolution of the spectrometre can also affect the quality of the LIBS measurement, since high resolution systems can separate spectral emission lines in close juxtaposition, reducing interference and increasing selectivity. This feature is particularly important in specimens which have a complex matrix, containing a large number of different elements. Accompanying the spectrometre and detector is a delay generator which accurately gates the detector's response time, allowing temporal resolution of the spectrum.

Advantages

Because such a small amount of material is consumed during the LIBS process the technique is considered essentially non-destructive or minimally-destructive, and with an average power density of less than one watt radiated onto the specimen there is almost no specimen heating surrounding the ablation site. Due to the nature of this technique sample preparation is typically minimised to homogenisation or is often unnecessary where heterogeneity is to be investigated or where a specimen is known to be sufficiently homogeneous, this reduces the possibility of contamination during chemical preparation steps. One of the major advantages of the LIBS technique is its ability to depth profile a specimen by repeatedly discharging the laser in the same position, effectively going deeper into the specimen with each shot. This can also be applied to the removal of surface contamination, where the laser is discharged a number of times prior to the analysing shot. LIBS is also a very rapid technique giving results within seconds, making it particularly useful for high volume analyses or on-line industrial monitoring.

LIBS is an entirely optical technique, therefore it requires only optical access to the specimen. This is of major significance as fibre optics can be employed for remote analyses. And being an optical technique it is non-invasive, non-contact and can even be used as a stand-off analytical technique when coupled to appropriate telescopic apparatus. These attributes have significance for use in areas from hazardous environments to space exploration. Additionally LIBS systems can easily be coupled to an optical microscope for micro-sampling adding a new dimension of analytical flexibility. With specialised optics or a mechanically positioned specimen stage the laser can be scanned over the surface of the specimen allowing spatially resolved chemical analysis and the creation of 'elemental maps'. This is very significant as chemical imaging is becoming more important in all branches of science and technology. Portable LIBS systems are more sensitive, faster and can detect a wider range of elements (particularly the light elements) than competing techniques such as portable x-ray fluorescence. And LIBS does not use ionizing radiation to excite the sample, which is both penetrating and potentially carcinogenic.

Disadvantages

LIBS, like all other analytical techniques is not without limitations. It is subject to variation in the laser spark and resultant plasma which often limits reproducibility. The accuracy of LIBS measurements is typically better than 10% and precision is often better than 5%. The detection limits for LIBS vary from one element to the next depending on the specimen type and the experimental apparatus used. Even so detection limits of 1 to 30 ppm by mass are not uncommon, but can range from >100 ppm to <1 ppm.

Recent Developments

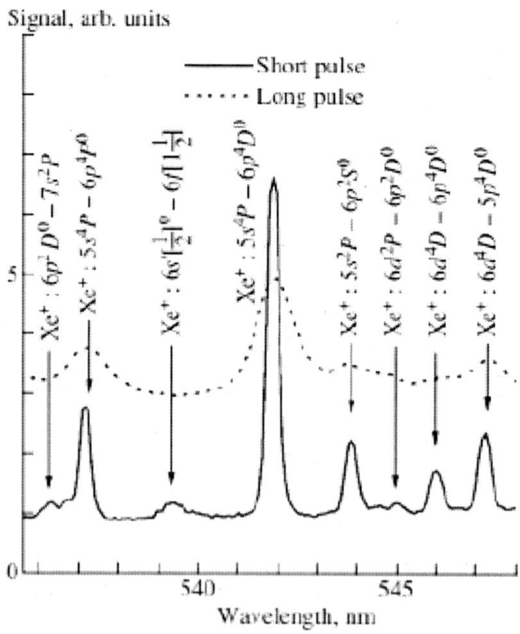

Figure: *. Spectrum of Xe at a pressure of 10 Torr using long pulses (5 ns, Nd:YAG laser) with an energy of 100 mJ/pulse and short pulse (200 fs, Ti:Sapphire laser) with an energy of 40 mJ/pulse. The strong line broadening in the case of long pulses is evident.*

Recent interest in LIBS has focused on the miniaturization of the components and the development of compact, low power, portable systems. This direction has been pushed along by interest from groups such as NASA, ESA as well as the military. The Mars Science Laboratory mission brought ChemCam, a LIBS instrument, onto Mars in 2012.

Recent developments in LIBS have seen the introduction of double-pulsed laser systems. For double-pulse LIBS one distinguishes between orthogonal and perpendicular configuration. In perpendicular configuration the laser is fired twice on the same spot on the specimen with a pulse separation in the order of one to a couple of tens of microseconds. Depending on pulse separation, the second pulse is more or less absorbed by the plasma plume caused by the previous pulse, resulting in a reheating of the laser plasma leading to signal enhancement. In orthogonal configuration a laser pulse is fired parallel to the sample surface either before or after the perpendicular pulse hits the specimen. The laser plasma ignited in the surrounding medium above the surface by a first pulse causes (by its shock wave) an area of reduced pressure above the specimen into which the actual plasma from the sample can expand. This has similar positive effects on sensitivity like LIBS performed at reduced pressures. If the orthogonal laser pulse is delayed with respect to the perpendicular one, the effects are similar as in the perpendicular configuration. Timing electronics such as digital delay generators can precisely control the timing of both pulses.

Both double-pulse LIBS as well as LIBS at reduced pressures are aimed at increasing the sensitivity of LIBS and the reduction of errors caused by the differential volatility of elements (e.g. Hydrogen as an impurity in solids). It also significantly reduces the matrix effects. Double-pulsed systems are also proving useful in conducting analysis in liquids, as the initial laser pulse forms a cavity bubble in which the second pulse acts on the evaporated material.

LIBS is one of several analytical techniques that can be deployed in the field as opposed to pure laboratory techniques e.g. spark OES. Recent research on LIBS is focusing on compact and (man-)portable systems. Industrial applications of LIBS are e.g. the detection of material mix-ups, analysis of inclusions in steel, analysis of slags in secondary metallurgy and high-speed identification of scrap pieces for material specific recycling tasks.

LIBS Using Short Laser Pulses

Following multiphoton or tunnel ionization the electron is being accelerated by inverse Bremsstrahlung and can collide with the nearby molecules and generate new electrons through collisions. If the pulse duration is long, the newly ionized electrons can be accelerated and eventually avalanche or cascade ionization follows. Once the density of the electrons reached a critical value breakdown occurs and high density plasma is created which has no memory of the laser pulse. So, the criterion for the shortness of a pulse in dense media is as the following: A pulse interacting with a dense matter is considered to be short if during the interaction the threshold for the avalanche ionization is not reached. At the first glance this definition may appear to be too limiting. Fortunately, due to the delicately balanced behaviour of the pulses in dense media the threshold does not reached easily. The phenomenon responsible for the balance is the intensity clamping during through the onset of filamentation process during the propagation of strong laser pulses in dense media.

A potentially important development to LIBS is the use of short laser pulse as a spectroscopic source. In this method, a plasma column is created as a result of focusing ultrafast laser pulses in a gas. The self-luminous plasma is far superior in terms of low level of continuum and also smaller line broadening. This is attributed to the lower density of the plasma in the case of short laser pulses due to the defocusing effects which limits the intensity of the pulse in the interaction region and thus prevents further multiphoton/tunnel ionization of the gas.

2

Laser Bonding

Laser bonding is a marking technique that uses lasers and other forms of radiant energy to bond an additive marking substance to a wide range of substrates.

Invented in 1997, this patent protected technology delivers permanent marks on metals, glass and ceramic parts for a diverse range of industrial and manual applications, ranging from aerospace to the awards & engraving industries. It differs from the more widely known techniques of laser engraving and laser ablation in that it is an additive process, adding material to the substrate to form the marking instead of removing it as in those techniques.

For metals, parts can be permanently marked with high contrast, high resolution marks for logos, bar-coding, identification and serialization purposes, without damage to the substrate. With glass and ceramics, complex surfaces can be decorated or marked and the traditional firing process replaced by a laser, with its permanent bond fired on in seconds.

Laser bonding has been achieved by Nd:YAG, CO_2 laser, Fibre laser and Diode-pumped solid-state laser and can be made by other forms of radiant energy.

Laser bonding is often referred to as "CerMark', "CerMarked", or "CerMarking". "CerMark" is a registered trademark of the Ferro Corporation, Cleveland Ohio, USA. This trademark, like Xerox, has become synonymous with the laser bonding process.

The Laser Bonding Process

Mark quality depends on a variety of factors, including the substrate used, marking speed, laser spot size, beam overlap, materials thickness, and laser parameters. Materials may be applied by various methods, including a brush on technique, spraying, pad printing, screen printing, roll coating, tape, and others.

The marking process generally comprises three steps:

1. Application of the marking material
2. Bonding of the marking material with a laser.
3. Removal of excess, unbonded material.

The resulting marking is fused to the substrate, and in most cases it is as durable as the substrate itself.

The Durability of Laser Bonded Markings

Markings placed on stainless steel are extremely durable and have survived such testing as abrasion resistance, chemical resistance, outdoor exposure, extreme heat, extreme cold, acids, bases and various organic solvents.

Marks on glass have been tested for resistance to acids, bases and scratching.

NASA's International Space Station, or ISS, was home to aluminum squares laser marked with CerMark marking material for almost four years. These squares were part of the Material International Space Station Experiment, or MISSE.

In this experiment test markings were applied to coupons made of materials commonly used in the construction of the external components used on space transportation vehicles, satellites and space stations. Markings applied using a wide range of different methods and techniques, including laser bonding. The material test coupons were then affixed to spaces provided on test panels, which were then installed onto trays which were attached to the ISS during a space walk conducted during the STS-105 Mission flown on August 10, 2001. The trays were positioned on the ISS so that they to could expect to receive the maximum amount of impact damage and exposure to a high degree of atomic oxygen and UV radiation.

The experiment was recovered on July 30, 2005 during STS-114 and returned to earth on August 9, 2005. The markings, DataMatrix two dimensional bar codes, were evaluated and found to be readable and visually looked as good as the day they were placed in orbit.

The laser bonding process is outlined and specified in both military and NASA marking specifications and standards. Laser bonding is also a preferred technique for use in the United States Department of Defense "Item Unique Identification" system (IUID).

Laser Engraving

Laser engraving, which is different than laser marking, is the practice of using lasers to engrave an object. Laser marking on the other hand just discolours the surface, without cutting into the surface. The technique does not involve the use of inks, nor does it involve tool bits which contact the engraving surface and wear out. These properties distinguish laser engraving from alternative engraving or marking technologies where inks or bit heads have to be replaced regularly.

The impact of laser marking has been more pronounced for specially designed "laserable" materials and also for some paints. These include laser-sensitive polymers and novel metal alloys.

The term laser marking is also used as a generic term covering a broad spectrum of surfacing techniques including printing, hot-branding and laser bonding. The machines for laser engraving and laser marking are the same, so that the two terms are sometimes confused by those without knowledge or experience, an honest mistake.

Laser Engraving Machines

A laser engraving machine can be thought of as three main parts: a laser, a controller, and a surface. The laser is like a pencil - the beam emitted from it allows the controller to trace patterns onto the surface. The controller (usually a computer) controls the direction, intensity, speed of movement, and spread of the laser beam aimed at the surface. The surface is picked to match what the laser can act on.

There are three main genres of engraving machines: The most common is the X-Y table where, usually, the workpiece (surface) is stationary and the laser optics move around in X and Y directions, directing the laser beam to draw vectors. Sometimes the laser is stationary and the workpiece moves. Sometimes the workpiece moves in the Y axis and the laser in the X axis. A second genre is for cylindrical workpieces (or flat workpieces mounted around a cylinder) where the laser effectively traverses a fine helix and on/off laser pulsing produces the desired image on a raster basis. In the third method, both the laser and workpiece are stationary and galvo mirrors move the laser beam over the workpiece surface. Laser engravers using this technology can work in either raster or vector mode.

The point where the laser (the terms “laser” and “laser beam” may be used interchangeably) touches the surface should be on the focal plane of the laser’s optical system, and is usually synonymous with its focal point. This point is typically small, perhaps less than a fraction of a millimetre (depending on the optical wavelength). Only the area inside this focal point is significantly affected when the laser beam passes over the surface. The energy delivered by the laser changes the surface of the material under the focal point. It may heat up the surface and subsequently vaporize the material, or perhaps the material may fracture (known as “glass” or “glass up”) and flake off the surface. Cutting through the paint of a metal part is generally how material is laser engraved.

If the surface material is vaporized during laser engraving, ventilation through the use of blowers or a vacuum pump are almost always required to remove the noxious fumes and smoke arising from this process, and for removal of debris on the surface to allow the laser to continue engraving.

A laser can remove material very efficiently because the laser beam can be designed to deliver energy to the surface in a manner which converts a high percentage of the light energy into heat. The beam is highly focused and collimated - in most non-reflective materials like wood, plastics and enamel surfaces, the conversion of light energy to heat is more than {x%} efficient. However, because of this efficiency, the equipment used in laser engraving may heat up rather quickly. Elaborate cooling systems are required for the laser. Alternatively, the laser beam may be pulsed to decrease the amount of excessive heating.

Different patterns can be engraved by programming the controller to traverse a particular path for the laser beam over time. The *trace* of the laser beam is carefully regulated to achieve a consistent removal depth of material. For example, criss-crossed paths are avoided to ensure that each etched surface is exposed to the laser only once, so the same amount of material is removed. The speed at which the beam moves across the material is also considered in creating engraving patterns. Changing the intensity and spread of the beam allows more flexibility in the design. For example, by changing the proportion of time (known as “duty-cycle”) the laser is turned on during each pulse, the power delivered to the engraving surface can be controlled appropriately for the material.

Since the position of the laser is known exactly by the controller, it is not necessary to add barriers to the surface to prevent the laser from deviating from the prescribed engraving pattern. As a result, no resistive mask is needed in laser engraving. This is primarily why this technique is different from older engraving methods.

A good example of where laser engraving technology has been adopted into the industry norm is the production line. In this particular setup, the laser beam is directed towards a rotating or vibrating mirror. The mirror moves in a manner which may trace out numbers and letters onto the surface being marked. This is particularly useful for printing dates, expiry codes, and lot numbering of products travelling along a production line. Laser marking allows materials made of plastic and glass to be marked "on the move". The location where the marking takes place is called a "marking laser station", an entity often found in packaging and bottling plants. Older, slower technologies such as hot stamping and pad printing have largely been phased out and replaced with laser engraving.

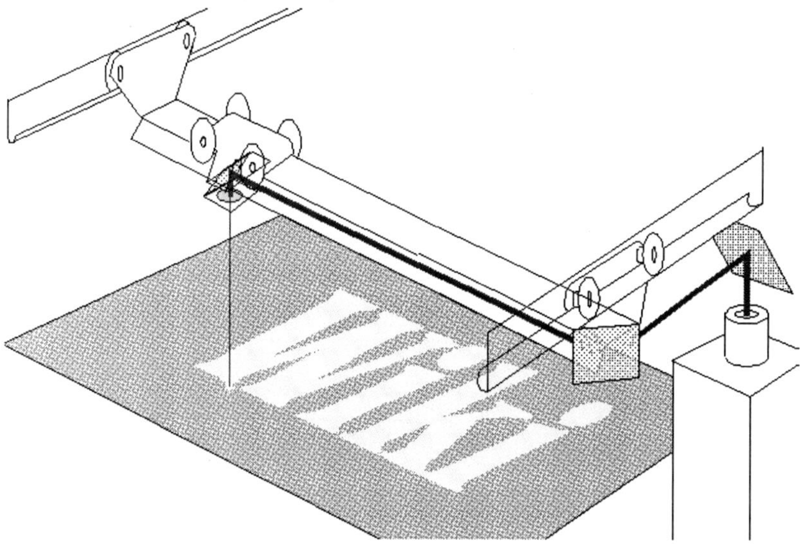

Figure: *Mirrors on both X and Y carriages allow exact positioning.*

For more precise and visually decorative engravings, a laser table is used. A laser table (or "X-Y table") is a sophisticated setup of equipment used to guide the laser beam more precisely. The laser is usually fixed permanently to the side of the table and emits light towards a pair of movable mirrors so that every point of the table surface can be swept by the laser. At the point of engraving, the laser beam is focused through a lens at the engraving surface, allowing very precise and intricate patterns to be traced out.

A typical setup of a laser table involves the fixed laser emitting light parallel to one axis of the table aimed at a mirror mounted on the end of an adjustable rail. The beam reflects off the mirror angled at 45 degrees so that the laser travels a path exactly along the length of the rail. This beam is then reflected by another mirror mounted to a movable trolley which directs the beam perpendicular to the original axis. In this scheme, two degrees of freedom (one vertical, and one horizontal) for etching can be represented.

In other laser engraving devices such as *flat table* or *drum engraving*, the laser beam is controlled to direct most of its energy a fixed penetration depth into the material to be engraved. In this manner, only a particular depth of material is removed when the engraving takes place. A simple machined stick or angle-iron can be used as a tool to help trained technologists adjust the engraver to achieve the required focusing. This setup is preferred for surfaces which do not vary in height appreciably.

For surfaces that vary in height, more elaborate focusing mechanisms have been developed. Some are known as *dynamic auto focus systems*. They adjust the lasing parameters in real time to adapt to the changes to the material as it is being etched. Typically, the height and depth of the surface is monitored with devices tracking changes to ultrasound, infrared, or visible light aimed at the engraving surface. These devices, known as *pilot beams* or *pilot lasers* (if a laser is used) help guide the adjustments made to the lens of the laser in determining the optimal spot to focus on the surface and remove material effectively.

"X-Y" laser engraving machines may operate in vector and raster mode.

Vector engraving follows the line and curve of the pattern to be engraved, much like a pen-based plotter draws by constructing line segments from a description of the outlines of a pattern. Much early engraving of signs and plaques (laser or otherwise) used pre-stored font outlines so that letters, numbers or even logos could be scaled to size and reproduced with exactly defined strokes. Unfortunately, "fill" areas were problematic, as cross-hatching patterns and dot-fills sometimes exhibited moiré effects or fibre-patterns caused by the imprecise calculation of dot spacings. Moreover, rotations of a font or dynamic scaling often were beyond the capabilities of the font-rendering device. The introduction of the PostScript page-description language now allows much greater flexibility—now virtually anything that can be described in vectors by PostScript-enabled software like CorelDRAW or Adobe Illustrator can be outlined, filled with suitable patterns, and laser-engraved.

Raster engraving traces the laser across the surface in a back-and-forth slowly advancing linear pattern that will remind one of the printhead on an inkjet or similar printer. The pattern is usually optimized by the controller/computer so that areas to either side of the pattern which aren't to be engraved are ignored and the trace across the material is thus shortened for better efficiency.

The amount of advance of each line is normally less than the actual dot-size of the laser; the engraved lines overlap just slightly to create a continuity of engravure. As is true of all rasterized devices, curves and diagonals can sometimes suffer if the length or position of the raster lines varies even slightly in relation to the adjacent raster scan; therefore exact positioning and repeatability are critically important to the design of the machine.

The advantage of rasterizing is the near effortless "fill" it produces. Most images to be engraved are bold letters or have large continuously engraved areas, and these are well-rasterized. Photos are rasterized (as in printing), with dots larger than that of the laser's spot, and these also are best engraved as a raster image.

Almost any page-layout software can be used to feed a raster driver for an X-Y or drum laser engraver. While traditional sign and plaque engraving tended to favour the solid strokes of vectors out of necessity, modern shops tend to run their laser engravers mostly in raster mode, reserving vector for a traditional outline "look" or for speedily marking out lines or "hatches" where a plate is to be cut.

Materials that can be Engraved

Natural Materials

Directly "burning" images on wood were some of the first uses of engraving lasers. The laser power required here is often less than 10 watts depending on the laser being used as most are different. Hardwoods like walnut, mahogany and maple produce good results. Softwoods can be judiciously engraved but tend to vaporize at less-consistent depths.

Burning a softwood with a fan blowing on it requires lowest power, quickest speed of cut, and enough airflow to extinguish what is trying meanwhile to ignite. Hard papers and fibreboard work well; linty papers and newsprint are like softwoods. Fur is not engraveable; finished leathers though can be laser-engraved with a look very similar to hot-branding. Certain latex rubber compounds can be laser engraved; for example these can be used to fabricate inking-stamps.

Paper masking tape is sometimes used as a pre-engraving overcoat on finished and resiny woods so that cleanup is a matter of picking the tape off and out of the unengraved areas, which is easier than removing the sticky and smoky surround "halos" (and requires no varnish-removing chemicals).

Plastics

Standard cast acrylic plastic, acrylic plastic sheet, and other cast resins generally laser very well. A commonly engraved award is a cast acrylic shape designed to be lasered from the back side. Styrene (as in compact disc cases) and many of the thermoforming plastics will tend to melt around the edge of the engraving spot.

The result is usually "soft" and has no "etch" contrast. The surface may actually deform or "ripple" at the lip areas. In some applications this is acceptable; for example date markings on 2-litre soda bottles do not need to be sharp.

For signage and face plates, etc., special laser-marked plastics were developed. These incorporate silicate or other materials which conduct excess heat away from the material before it can deform. Outer laminates of this material vaporize easily to expose different coloured material below.

Other plastics may be successfully engraved, but orderly experimentation on a sample piece is recommended. Bakelite is said to be easily laser-engraved; some hard engineering plastics work well. Expanded plastics, foams and vinyls however are generally candidates for routing rather than laser engraving. Plastics with a chlorine content (such as vinyl, PVC) produce corrosive chlorine gas when lasered, which combines with Hydrogen in the air to produce vaporized hydrochloric acid which can damage a laser engraving system. Urethane and silicone plastics usually don't work well—unless it is a formulation filled with cellulose, stone or some other stable insulator material.

Many light switchplates from companies such as Leviton or Lutron can be laser engraved. Again, experimentation may be necessary to develop the correct laser settings to result in engraving the surface rather than melting it. Often the laser engraving is followed by back filling with paint on the engraved surface to produce more contrast between the engraved surface and the surrounding surface.

Kevlar can be laser-engraved and laser-cut. However, Kevlar does give off extremely hazardous fumes (cyanide gas) when it is vaporized.

Metals

The best traditional engraving materials started out being the worst laser-engravable materials. This problem has now been solved using lasers at shorter wavelengths than the traditional 10,640 nm wavelength CO_2 laser. Using Yb: Fibre Lasers, Nd:YVO_4 or Nd:YAG lasers at 1,064 nm wavelength, or its harmonics at 532 and 355 nm, metals can now easily be engraved using commercial systems.

***Figure:** Laser on Stainless Steel (SS316L)*

Coated Metals

The same conduction that works against the spot vaporization of metal is an asset if the objective is to vaporize some other coating away from the metal. Laser engraving metal plates are manufactured with a finely polished metal, coated with an enamel paint made to be "burned off". At levels of 10-30 watts, excellent engravings are made as the enamel is removed quite cleanly. Much laser engraving is sold as exposed brass or silver-coated steel lettering on a black or dark-enamelled background. A wide variety of finishes are now available, including screen-printed marble effects on the enamel.

Spray coatings can be obtained for the specific use of laser engraving metals, these sprays apply a coating that is visible to the laser light which fuses the coating to the substrate where the laser passed over. Typically, these sprays can also be used to engrave other optically invisible or reflective substances such as glass and are available in a variety of colours. Besides spray coatings, some laser-markable metals come pre-coated for imaging. Products such as this transform the surface of the metal to a different colour (often black, brown or grey).

Stone and Glass

Stone and glass do not turn gaseous very easily. As expected, this makes them generally a better candidate for other means of engraving, most notably sandblasting or cutting using diamonds and water. But when a laser hits glass or stone, something else interesting happens: it fractures. Pores in the surface expose natural grains and crystalline "stubs" which, when heated very quickly, can separate a microscopic sized "chip" from the surface because the hot piece is expanding relative to its surroundings. So lasers are indeed used to engrave on glass, and if the power, speed and focus are just right, excellent results can be achieved. One should avoid large "fill" areas in glass engraving because the results across an expanse tend to be uneven; the glass ablation simply cannot be depended on for visual consistency, which may be a disadvantage or an advantage depending on the circumstances and the desired effect.

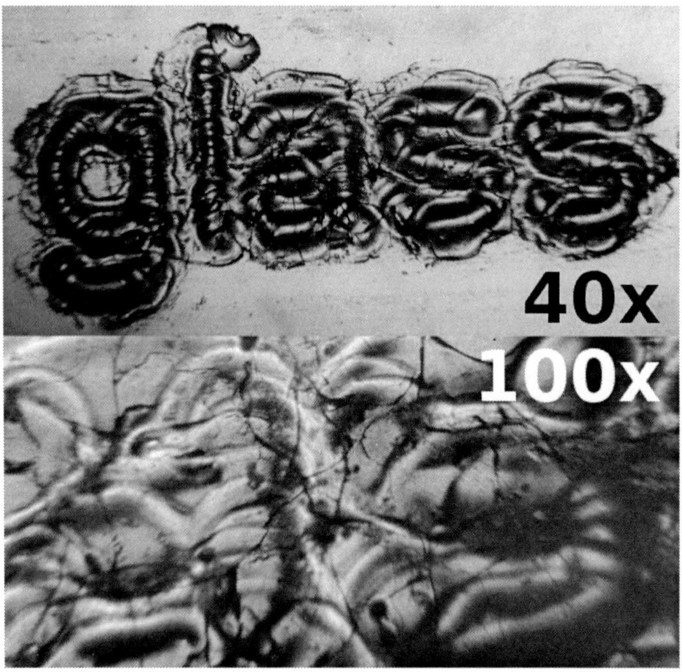

Figure: *Laser engraved glass microscope slide with the word "glass" engraved in 3pt font. Magnified to 40x and 100x*

Jewellery

The demand for personalized jewellery has made jewellers more aware of the benefits of the laser engraving process.

Jewellers found that by using a laser, they could tackle an engraving task with greater precision. In fact, jewelers discovered that laser engraving allowed for more precision than other types of engraving. At the same time, jewellers discovered that laser applied engravings had a number of other desirable features.

At one time jewellers who attempted to do laser engraving did need to use large pieces of equipment. Now the devices that perform laser engraving come in units. Some entrepreneurs have placed such units in mall kiosks. That has made laser engraving jewellery much more accessible. The makers of machines for laser engraving jewellers have developed some very specialised equipment. They have designed machines that can engrave the inside of a ring. They have also created machines that have the ability to engrave the back of a watch.

A laser can cut into both flat and curved surfaces such as the surfaces on jewellery. That points out the reason why jewellers have welcomed all the adaptations for the creation of laser engraved jewellery.

Fine Art

Laser engraving can also be used to create works of fine art. Generally this involves engraving into planar surfaces, to reveal lower levels of the surface or to create grooves and striations which can be filled with inks, glazes, or other materials. Some laser engravers have rotary attachments which can engrave around an object. Artists may digitize drawings, scan or create images on a computer, and engrave the image onto any of the materials cited in this article.

Industrial Applications

Direct Laser Engraving of Flexographic Plates and Cylinders

Direct laser engraving of flexographic printing cylinders and plates has been an established process since the 1970s. This first began with the use of a carbon dioxide laser used to selectively ablate or evaporate a variety of rubber plate and sleeve materials to produce a print ready surface without the use of photography or chemicals. With this process there is no integral ablation mask as with direct photopolymer laser imaging. Instead a high-power carbon dioxide laser head burns away, or ablates, unwanted material. The aim is to form sharp, relief images with steep first relief and contoured shoulder supported edges to give a high standard of process colour reproduction. A short water wash and dry cycle follows, which is a lot less involved than in the post-processing stages for direct laser imaging or conventional flexo platemaking using photopolymer plates. After engraving, the photopolymer is exposed through the imaged black layer and washed out in the traditional photopolymer process requiring photography and chemicals.

Before the year 2000 lasers only produced lower quality in rubber-like materials. In these rubber-like materials, which had a rough structure, higher quality was impossible. Since the year 2000 fibre lasers have been introduced to give a much increased engraving quality direct into black polymeric materials. Also at the Drupa 2004 the direct engraving of polymer plates was introduced. This had also an effect on the rubber-developers who, in order to stay competitive, developed new high quality rubber-like materials. The development of suitable polymeric compounds has also allowed the engraving quality achievable with the fibre lasers to be realised in print. Since then direct laser engraving of flexo-printingforms is seen by many as the modern way to make printing-forms for it is the first real digital way.

As a competitive process, more recently laser system have been introduced to selectively engrave the thin opaque black layer of a specially produced photopolymer plate or sleeve.

Direct Photopolymer Laser Imaging

Closely related is the direct imaging of a digital flexo plates or sleeves 'in-the-round' on a fast-rotating drum, or cylinder. This is carried out on a platesetter integrated within a digital prepress workflow, that also supports digital proofing. Again, this is a filmless process, which removes one of the variables in obtaining the fine and sharp dots for screened affects, including process colour printing.

With this process the electronically generated image is scanned at speed to a photopolymer plate material that carries a thin black mask layer on the surface. The infrared laser-imaging head, which runs parallel to the drum axis, ablates the integral mask to reveal the uncured polymer underneath. A main ultraviolet exposure follows to form the image through the mask. The remaining black layer absorbs the ultraviolet radiation, which polymerizes the underlying photopolymer where the black layer has been removed. The exposed digital plate still needs to be processed like a conventional flexo plate. That is, using solvent-based washout with the necessary waste recovery techniques, although some water-washable digital plates are in development. This technology has been used since 1995 and is only now becoming more widely used around the world as more affordable equipment becomes available. Trade sources say there are around 650 digital platesetters installed in label, packaging and trade platemaking houses.

Laser Engraving of Anilox Rolls

Prior to 1980 anilox rolls were produced by a variety of mechanical processes. These metal anilox rolls were sometimes sprayed with ceramic to prolong their life in the flexographic printing press. During the 1980s laser engraving systems were produced which used a carbon dioxide laser to engrave the required cell pattern directly into the polished ceramic surface. Since then Q-switched YAG lasers were used for a period as they provided a more focusable laser beam as well as increased pulsing frequencies capable of engraving the finer cell configuration demanded by the ever evolving flexographic printing process.

Since approximately the year 2000 the direct anilox laser engraving process has been dominated by the use of fibre lasers which provide the high powers of the carbondioxide lasers together with the finely focusable beam of the YAG lasers.

Optical systems providing the rapid switching of multiple beams have allowed the fibre laser system to be dominant in this market. This technology has become known as Multi-Beam-Anilox or MBA.

Sub-Surface Laser Engraving (SSLE)

Sub-surface laser engraving is the process of engraving an image in a transparent solid material by focusing a laser below the surface to create small fractures. Such engraved materials are of high-grade optical quality (suitable for lenses, with low dispersion) to minimize distortion of the beam. BK7 glass is a common material for this application. Plastics are also used.

Since its inception in the late 1990s, SSLE has become more cost effective with a number of different sized machines ranging from small (~US$35,000–60,000) to large production sized tables (>US$250,000). Although these machines are becoming more available, it is estimated that only a few hundred are in operation worldwide. Many machines require very expensive cooling, maintenance and calibration for proper use. The more popular SSLE engraving machines use the Diode Pumped Solid State or DPSS laser process.

The laser diode, the primary component which excites a pulsed solid state laser, can easily cost one third of the machine itself and functions for a limited number of hours, although a good quality diode can last thousands of hours.

In the past five years, use of SSLE has become more cost effective to produce 3D images in souvenir 'crystal' or promotional items with only a few designers concentrating on designs incorporating large or monolithic sized crystal.

A number of companies offer custom made souvenirs by taking 3D pictures or photos and engraving them into the crystal. Quality of the designs and images varies greatly between vendors in the promotional and personal services sector (photo engravers) – the mass-producing curio vendors have the habit of reducing resolution of the points and lowering output to maximize their laser diode's lifespan.

Optical Breakdown Photoionization Mode (OB) at Photoionization Mode

A photoionization mode is a mode of interaction between a laser beam and matter involving photoionization.

Laser light affects materials of all types through fundamental processes such as excitation, ionization, and dissociation of atoms and molecules. These processes depend on the proprieties of the light, as well as on the proprieties of the material. Using lasers for material processing requires understanding and being able to control these fundamental effects. A better understanding can be achieved by defining distinct interaction regimes, hence the definition of four photoionization modes.

This new way of looking at the laser interaction with matter was first proposed by Tiberius Brastaviceanu in 2006, after his description of the "filamentary ionization mode" (Sherbrooke University, 2005). In his Master's work he provided the empirical proof of the formation of filamentary distributions of solvated electrons in water, induced by high-power fs (femtosecond, one trillionth of a second) laser pulses in the self-focusing propagation regime, and described the theoretical context in which this phenomenon can be explained and controlled.

Single-Photon Photoionization Mode (SP)

The SP mode is obtained at small wavelengths (UV, X-ray), or high energy per photon, and at low intensity levels. The only photoionization process involved in this case is the single-photon ionization.

Optical Breakdown Photoionization Mode (OB)

The OB mode is observed when a material is subjected to very powerful laser pulses. It manifests a power threshold in the range of MW for the majority of dielectric materials, which depends on the duration and on the wavelength of the laser pulse. Optical breakdown is related to the dielectric breakdown phenomenon which was studied and modelled successfully towards the end of the 1950s. One describes the effect as a strong local ionization of the medium, where the plasma reaches densities beyond the critical value (between 10^{20} and 10^{22} electrons/cm^3). Once the plasma critical density is achieved, energy is very efficiently absorbed from the light pulse, and the local plasma temperature increases dramatically. An explosive Coulombian expansion follows, and forms a very powerful and damaging shockwave through the material that develops on ns timescale. In liquids one talks about cavitation bubbles. If the rate of plasma formation is relatively slow, in the nanosecond time regime (for nanosecond excitation laser pulses), energy is transferred from the plasma to the lattice, and thermal damages can occur. In the femtosecond time regime (for femtosecond excitation laser pulses) the plasma expansion happens on a timescale smaller than the rate of energy transfer to the lattice, and thermal damages are reduced or eliminated. This is the bases if cold laser machining using high-power sub-ps laser sources.

The optical breakdown is a very "violent" phenomenon and changes drastically the structure of the surrounding medium. To the naked eye, optical breakdown looks like a spark and if the event happens in air or some other fluid, it is even possible to hear a short noise (burst) caused by the explosive plasma expansion.

There are several photoionization processes involved in optical breakdown, which depend on the wavelength, local intensity, and pulse duration, as well as on the electronic structure of the material. First, we should mention that optical breakdown is only observed at very high intensities. For pulse durations greater than a few tens of fs avalanche ionization plays a role. The longer pulse duration, the greater the avalanche ionization's contribution. Multi-photon ionization processes are important in the fs time regime, and their role increases as the pulse duration decreases. The type of multi-photon ionization processes involved is also wavelength dependent.

The theory needed to understand the most important features of optical breakdown are:

- the physics of strong-(laser)field interaction with matter, to account for the plasma formation;
- the physics of strong-(laser)field interaction with plasma, to account for plasma expansion, and for thermal and mechanical effects;
- the geometrical/linear optical theory, to account at the first approximation for the spatial intensity distribution. Non-linear propagation theory is usually invoked to account for self-focusing that occurs in experiments conducted at low numerical aperture, and to account for detail features of the plasma density spatial distribution.

Below Optical Breakdown Threshold Photoionization Mode (B/OB)

B/OB mode is an intermediary between the optical breakdown mode (OB mode) and the filamentary mode (F mode). The plasma density generated in this mode can go from 0 to the critical value i.e. optical breakdown threshold. Intensities reached inside the B/OB zone can range from multi-photon ionization threshold to the optical breakdown threshold. In the visible-IR domain, B/OB mode is obtained under very tight external focusing (high numerical aperture), to avoid self-focusing, and for intensities below optical breakdown threshold. In the UV regime, where optical breakdown intensity threshold is below self-focusing intensity threshold, tight focusing is not necessary. The shape of the ionization area is similar to that of the focal area of the beam, and can be very small (only a few micrometres). B/OB mode is possible only at short pulse durations, where AI's contribution to the total free electron population is very small. As the pulse duration becomes even shorter, the intensity domain where B/OB is possible becomes even wider.

The principles governing this mode of ionization are very simple. Localized plasma must be generated in predictable fashion, under the optical breakdown threshold. Optical breakdown intensity threshold is strongly correlated to the input intensity only at short pulse durations. Therefore, one important requirement, in order to systematically avoid the optical breakdown, is to operate at short pulse durations. In order for the ionization to take place, multi-photon ionization (MPI)intensity threshold must be reached. The idea is to adjust the duration of the laser pulse so that multi-photon ionization, and perhaps to a lesser extent avalanche ionization, have no time to raise the plasma's density above the critical value. In the UV, the distinction between single-photon mode (SP) and B/OB is that for the latter multi-photon ionization, single-photon ionization, and perhaps to a lesser extent avalanche ionization, are operating, whereas for the former, only single-photon ionization is operating.

B/OB relies mostly on MPI processes. Therefore, it is more selective than OB in terms of which type of atom or molecule is ionized or dissociated. The theory needed to understand the most important features of B/OB are:

- The physics of strong-(laser)field interaction with matter, to account for the plasma formation. As opposed to the OB mode, in this case the role of avalanche ionization is greatly reduced, and the effects are dominated by multi-photon ionization processes.
- The geometrical/linear optical theory, to account at the first approximation for the spatial intensity distribution. Non-linear propagation theory is usually invoked to account for self-focusing that occurs in experiments conducted at low numerical aperture, and to account for detailed features of plasma spatial distribution.

The B/OB mode was described by A. Vogel et al. [ref 2].

Filamentary Photoionization Mode (F)

In the F mode, filamentary or linear ionization patterns are formed. The plasma density within these filaments is below the critical value.

The self-focusing effect is responsible for the most important characteristics of the dose distribution. The diameter of these filamentary ionization traces is the same within 20% (in the order of a few micrometres). Their length, their number, and their relative position are controllable parameters. The plasma density and the yield of photolytic species are believed to be homogeneously distributed along these filaments. The local intensity reached by the laser light during propagation is also practically constant along their length. The power range of operation of the F mode is above self-focusing threshold and below optical breakdown threshold. Consequently, a necessary condition for it to exist is that the self-focusing threshold must be smaller than the optical breakdown threshold.

The F mode exhibits very important characteristics, which in combination with the other three photoionization modes makes possible the generation of a wide range of dose distributions, expanding the application range of lasers in the domain of material processing. The F mode is the only mode capable of generating linear ionization traces.

The theory needed to understand the most important features of the F mode are:

- The physics of high-(laser)field interaction with matter, to account for the plasma formation
- The theory of non-linear propagation, to account for the spatial redistribution of the laser light, intensity clamping, and the formation of filaments, as well as for frequency conversion processes.

The first concrete connection between non-linear optical effects, such as the supercontinuum generation, and photoionization was established by A. Brodeur and S.L. Chin [ref 4] in 1999, based on optical experimental data and modelling. In 2002 T. Brastaviceanu published the first direct measurement of the spatial distribution of photoionization induced in the self-focusing regime, in water [ref 5].

Superposition of Photoionization Modes

It is possible to control the spatial distribution of the dose induced by laser pulses, and the relative yields of primary photolytic species, by controlling the properties of the laser beam. The dose distribution can be conveniently shaped by inducing a superposition of the four modes of photoionization. The mixed ionization modes are: SP-OB, SP-B/OB, and F-OB.

Soft Retooling

Soft retooling is a process that changes the ablation shape during laser ablation without changing any hardware. Contrarily to conventional milling, it is possible with laser sculpting to change ablation geometry just by changing parameters of laser ablation that have the desired effects on laser ablation. Conventional milling machines contain trays on which different tool bits are stored and must consider in their machining routine tool change. This operation is not necessary when using laser sculpting.

One-Dimensional Metal-Oxide Nanostructures: Synthesis, Characterization, and Applications

An oxide /ÈRksajd/ is a chemical compound that contains at least one oxygen atom and one other element in its chemical formula. Metal oxides typically contain an anion of oxygen in the oxidation state of –2. Most of the Earth's crust consists of solid oxides, the result of elements being oxidized by the oxygen in air or in water. Hydrocarbon combustion affords the two principal carbon oxides: carbon monoxide and carbon dioxide. Even materials considered pure elements often develop an oxide coating. For example, aluminium foil develops a thin skin of Al_2O_3 (called a passivation layer) that protects the foil from further corrosion. Different oxides of the same element are distinguished by Roman numerals denoting their oxidation number, e.g. iron(II) oxide versus iron(III) oxide.

Formation

Due to its electronegativity, oxygen forms stable chemical bonds with almost all elements to give the corresponding oxides. Noble metals (such as gold or platinum) are prized because they resist direct chemical combination with oxygen, and substances like gold(III) oxide must be generated by indirect routes.

Two independent pathways for corrosion of elements are hydrolysis and oxidation by oxygen. The combination of water and oxygen is even more corrosive. Virtually all elements burn in an atmosphere of oxygen, or an oxygen rich environment. In the presence of water and oxygen (or simply air), some elements— sodium—react rapidly, even dangerously, to give the hydroxides. In part for this reason, alkali and alkaline earth metals are not found in nature in their metallic, i.e., native, form. Caesium is so reactive with oxygen that it is used as a getter in vacuum tubes, and solutions of potassium and sodium, so-called NaK are used to deoxygenate and dehydrate some organic solvents. The surface of most metals consists of oxides and hydroxides in the presence of air. A well-known example is aluminium foil, which is coated with a thin film of aluminium oxide that passivates the metal, slowing further corrosion. The aluminium oxide layer can be built to greater thickness by the process of electrolytic anodising. Though solid magnesium and aluminium react slowly with oxygen at STP—they, like most metals, burn in air, generating very high temperatures. Finely grained powders of most metals can be dangerously explosive in air. Consequently, they are often used in Solid-fuel rockets.

In dry oxygen, iron readily forms iron(II) oxide, but the formation of the hydrated ferric oxides, $Fe_2O_{3-x}(OH)_{2x}$, that mainly comprise rust, typically requires oxygen *and* water. Free oxygen production by photosynthetic bacteria some 3.5 billion years ago precipitated iron out of solution in the oceans as Fe_2O_3 in the economically important iron ore hematite.

Structure

Oxides of most metals adopt polymeric structures with M-O-M crosslinks. Because these crosslinks are strong, the solids tend to be insoluble in solvents, though they are attacked by acids and bases. The formulas are often deceptively simple. Many are nonstoichiometric compounds. In these oxides, the coordination number of the oxide ligand is two for most electronegative elements and 3–6 for most metals.

Figure: *The unit cell of rutile. Ti(IV) centres are grey; oxide centres are red. Notice that oxide forms three bonds to titanium and titanium forms six bonds to oxide.*

Molecular Oxides

Although most metal oxides are polymeric, some oxides are molecules. The most famous molecular oxides are carbon dioxide and carbon monoxide.

Phosphorus pentoxide is a more complex molecular oxide with a deceptive name, the formula being P_4O_{10}. Some polymeric oxides when heated depolymerize to give molecules, examples being selenium dioxide and sulfur trioxide. Tetroxides are rare, and there are only five known examples: ruthenium tetroxide, osmium tetroxide, hassium tetroxide, iridium tetroxide, and xenon tetroxide.

Many oxyanions are known, such as polyphosphates and polyoxometalates. Oxycations are rarer, an example being nitrosonium (NO^+). Of course many compounds are known with both oxides and other groups. In organic chemistry, these include ketones and many related carbonyl compounds. For the transition metals, many oxo complexes are known as well as oxyhalides.

Reactivity

Oxides can be attacked by acids and bases. Those attacked only by acids are basic oxides; those attacked only by bases are acidic oxides. Oxides that react with both acids and bases are amphoteric. Metals tend to form basic oxides, non-metals tend to form acidic oxides, and amphoteric oxides are formed by elements near the boundary between metals and non-metals (metalloids).

This reactivity is the basis of many practical processes such, as the extraction of some metals from their ores in the process called hydrometallurgy.

Reduction

Metals are "won" from their oxides by chemical reduction. A common and cheap reducing agent is carbon in the form of coke. The most prominent example is that of iron ore smelting. Many reactions are involved, but the simplified equation is usually shown as:

$$2\ Fe_2O_3 + 3\ C \rightarrow 4\ Fe + 3\ CO_2$$

Metal oxides can be reduced by organic compounds. This redox process is the basis for many important transformations in chemistry, such as the detoxification of drugs by the P450 enzymes and the production of ethylene oxide, which is converted to antifreeze. In such systems the metal centre transfers an oxide ligand to the organic compound followed by regeneration of the metal oxide, often by oxygen in air.

Carbothermic Reaction

Carbothermic reactions involve the reduction of substances, often metal oxides, using carbon as the reducing agent. These chemical reactions are usually conducted at temperatures of several hundred degrees Celsius. Such processes are applied for production of the elemental forms of many elements. Carbothermic reactions are not however useful for some metal oxides such as those of sodium and potassium. The ability of metals to participate in carbothermic reactions can be predicted from Ellingham diagrams.

Carbothermal reactions produce carbon monoxide and sometimes carbon dioxide. The facility of these conversions is attributable to the entropy of reaction: two solids, the metal oxide and carbon, are converted to a new solid (metal) and a gas (CO), the latter having high entropy. Heat is required for carbothermic reactions because diffusion of the reacting solids is slow otherwise.

The most prominent example is that of iron ore smelting. Many reactions are involved, but the simplified equation is usually shown as:

$$2Fe2O3 + 3C \rightarrow 4Fe + 3CO_2$$

On a more modest scale, about 1 million tons of elemental phosphorus is produced annually by carbothermic reactions. Calcium phosphate (phosphate rock) is heated to 1,200–1,500 °C with sand, which is mostly SiO_2, and coke (impure carbon) to produce P_4. The chemical equation for this process when starting with fluoroapatite, a common phosphate mineral, is:

$$4Ca_5(PO_4)_3F + 18SiO_2 + 30C \rightarrow 3P_4 + 30CO + 18CaSiO_3 + 2CaF_2$$

Sometimes carbothermic reactions are coupled to other conversions. One example is the chloride process for separating titanium from ilmenite, the main ore of titanium. In this process, a mixture of carbon and the crushed ore is heated at 1000 °C under flowing chlorine gas, giving titanium tetrachloride:

$$2FeTiO_3 + 7Cl_2 + 6C \rightarrow 2TiCl_4 + 2FeCl_3 + 6CO$$

For some metals, carbothermic reactions do not afford the metal, but instead give the metal carbide. This behaviour is observed for titanium, hence the use of the chloride process. Carbides also form upon high temperature treatment of Cr_2O_3 with carbon. For this reason, aluminium is employed as the reducing agent.

Hydrolysis

Oxides of more electropositive elements tend to be basic. They are called *basic anhydrides*. Exposed to water, they may form basic hydroxides. For example, sodium oxide is basic—when hydrated, it forms sodium hydroxide. Oxides of more electronegative elements tend to be acidic. They are called "acid anhydrides"; adding water, they form oxoacids. For example, dichlorine heptoxide is acid; perchloric acid is a more hydrated form. Some oxides can act as both acid and base. They are amphoteric. An example is aluminium oxide. Some oxides do not show behaviour as either acid or base.

The oxide ion has the formula O^{2-}. It is the conjugate base of the hydroxide ion, OH^-, and is encountered in ionic solid such as calcium oxide. O^{2-} is unstable in aqueous solution " its affinity for H^+ is so great ($pK_b \sim -38$) that it abstracts a proton from a solvent H_2O molecule:

$$O^{2-} + H_2O \rightarrow 2\ OH^-$$

The equilibrium constant of aforesaid reactions is $pK_{eq} \sim -22$

In the 18th century, oxides were named calxes or calces after the calcination process used to produce oxides. *Calx* was later replaced by *oxyd*.

Nomenclature and Formulas

Sometimes, metal-oxygen ratios are used to name oxides. Thus, NbO would be called niobium monoxide and TiO_2 is titanium dioxide. This naming follows the Greek numerical prefixes. In the older literature and continuing in industry, oxides are named by contracting the element name with "a." Hence alumina, magnesia, chromia, are, respectively, Al_2O_3, MgO, Cr_2O_3.

Special types of oxides are peroxide, O_2^{2-}, and superoxide, O_2^-. In such species, oxygen is assigned higher oxidation states than oxide.

The chemical formulas of the oxides of the chemical elements in their highest oxidation state are predictable and are derived from the number of valence electrons for that element. Even the chemical formula of O_4, tetraoxygen, is predictable as a group 16 element. One exception is copper, for which the highest oxidation state oxide is copper(II) oxide and not copper(I) oxide. Another exception is fluoride, which does not exist as one might expect—as F_2O_7—but as OF_2.

Since fluorine is more electronegative than oxygen, oxygen difluoride (OF_2) does not represent an oxide of fluorine, but instead represents a fluoride of oxygen.

One of the biggest challenges of 21st century is to develop powerful electrochemical energy devices (EEDs). The EEDs such as fuel cells, supercapacitors, and Li-ion batteries are among the most promising candidates in terms of power-densities and energy-densities. The nanostructured materials (NSMs) have drawn intense attention to develop highly efficient EEDs because of their high surface area, novel size effects, significantly enhanced kinetics, and so on.

The main types of nanostructured materials based on the dimensions of their structural elements are: zero-dimensional (0D), one-dimensional (1D), two-dimensional (2D) and three-dimensional (3D) nanomaterials. Zero-dimensional nanomaterials include nanocluster materials and nanodispersions, i.e. materials in which nanoparticles are isolated from each other. One-dimensional nanomaterials are nanofibre (nanorod) and nanotubular materials with fibre (rod, tube) length from 100 nm to tens of microns. Two-dimensional nanomaterials are films (coatings) with nanometre thickness. Structural elements in 0D, 1D and 2D nanomaterials can be distributed in a liquid or solid macroscopic matrix or be applied on a substrate. Three-dimensional nanomaterials include powders, fibrous, multilayer and polycrystalline materials in which the 0D, 1D and 2D structural elements are in close contact with each other and form interfaces. An important type of three-dimensional nanostructured materials is a compact or consolidated (bulk) polycrystal with nanosize grains, whose entire volume is filled with those nanograins, free surface of the grains is practically absent, and there are only grain interfaces. The formation of such interfaces and "disappearance" of the nanoparticle (nanograin) surface is the fundamental difference between three-dimensional compact nanomaterials and nanocrystalline powders with various degrees of agglomeration that consist of particles of the same size as the compact nanostructured materials.

One dimensional nanomaterials have attracted much attention from academia and industry. A large series of devices have been created on the basis of 1D nanomaterials. Their applications have changed and will continuously change our daily life.

One-dimensional (1D) metal-oxide nanostructures are ideal systems for exploring a large number of novel phenomena at the nanoscale and investigating size and dimensionality dependence of nanostructure properties for potential applications. The construction and integration of photodetectors or optical switches based on such nanostructures with tailored geometries have rapidly advanced in recent years. Active 1D nanostructure photodetector elements can be configured either as resistors whose conductions are altered by a charge-transfer process or as field-effect transistors (FET) whose properties can be controlled by applying appropriate potentials onto the gates. Functionalizing the structure surfaces offers another avenue for expanding the sensor capabilities.

ZnO

Zinc oxide is an inorganic compound with the formula ZnO. ZnO is a white powder that is insoluble in water, and it is widely used as an additive in numerous materials and products including rubbers, plastics, ceramics, glass, cement, lubricants, paints, ointments, adhesives, sealants, pigments, foods (source of Zn nutrient), batteries, ferrites, fire retardants, and first-aid tapes. It occurs naturally as the mineral zincite, but most zinc oxide is produced synthetically.

Nanostructures of ZnO can be synthesized into a variety of morphologies including nanowires, nanorods, tetrapods, nanobelts, nanoflowers, nanoparticles etc. Nanostructures can be obtained with most above-mentioned techniques, at certain conditions, and also with the vapor-liquid-solid method.

Rodlike nanostructures of ZnO can be produced via aqueous methods, which are attractive for the following reasons: They are low cost, less hazardous, and thus capable of easy scaling up; the growth occurs at a relatively low temperature, compatible with flexible organic substrates; there is no need for the use of metal catalysts, and thus it can be integrated with well-developed silicon technologies. In addition, there are a variety of parameters that can be tuned to effectively control the morphology and properties of the final product. Wet chemical methods have been demonstrated as a very powerful and versatile technique for growing one-dimensional ZnO nanostructures.

The synthesis is typically carried out at temperatures of about 90 °C, in an equimolar aqueous solution of zinc nitrate and hexamine, the latter providing the basic environment. Certain additives, such as polyethylene glycol or polyethylenimine, can improve the aspect ratio of the ZnO nanowires. Doping of the ZnO nanowires has been achieved by adding other metal nitrates to the growth solution. The morphology of the resulting nanostructures can be tuned by changing the parameters relating to the precursor composition (such as the zinc concentration and pH) or to the thermal treatment (such as the temperature and heating rate).

Aligned ZnO nanowires on pre-seeded silicon, glass and gallium nitride substrates have been grown in aqueous solutions using aqueous zinc salts such as Zinc nitrate and Zinc acetate in basic environments. Pre-seeding substrates with ZnO creates sites for homogeneous nucleation of ZnO crystal during the synthesis. Common pre-seeding methods include in-situ thermal decomposition of zinc acetate crystallites, spincoating of ZnO nanoparticles and the use of physical vapor deposition methods to deposit ZnO thin films. Pre-seeding can be performed in conjunction with top down patterning methods such as electron beam lithography and nanosphere lithography to designate nucleation sites prior to growth. Aligned ZnO nanowires can be used in dye-sensitized solar cells and field emission devices.

SnO_2

Tin dioxide, also known by the systematic name tin(IV) oxide and stannic oxide in the older notation, is the inorganic compound with the formula SnO_2. The mineral form of SnO_2 is called cassiterite, and this is the main ore of tin. With many other names, this oxide of tin is the most important raw material in tin chemistry. This colourless, diamagnetic solid is amphoteric.

Structure

It crystallises with the rutile structure, wherein the tin atoms are six coordinate and the oxygen atoms three coordinate. SnO_2 is usually regarded as an oxygen-deficient n-type semiconductor. Hydrous forms of SnO_2 have been described in the past as stannic acids, although such materials appear to be hydrated particles of SnO_2 where the composition reflects the particle size.

Preparation

Tin dioxide occurs naturally but is purified by reduction to the metal followed by burning tin in air. Annual production is in the range of 10 kilotons. SnO_2 is reduced industrially to the metal with carbon in a reverbatory furnace at 1200-1300 °C.

Amphoterism

Although SnO_2 is insoluble in water, it is an amphoteric oxide, although cassiterite ore has been described as difficult to dissolve in acids and alkalis. "Stannic acid" refers to hydrated tin dioxide, SnO_2, which is also called "stannic hydroxide."

Tin oxides dissolve in acids. Halogen acids attack SnO_2 to give hexahalostannates, such as $[SnI_6]^{2-}$. One report describes reacting a sample in refluxing HI for many hours.

$$SnO_2 + 6\ HI \rightarrow H_2SnI_6 + 2\ H_2O$$

Similarly, SnO_2 dissolves in sulfuric acid to give the sulfate:

$$SnO_2 + 2\ H_2SO_4 \rightarrow Sn(SO_4)_2 + 2\ H_2O$$

SnO_2 dissolves in strong base to give "stannates," with the nominal formula Na_2SnO_3. Dissolving the solidified SnO_2/NaOH melt in water gives $Na_2[Sn(OH)_6]_2$, "preparing salt," which is used in the dye industry.

Uses

In conjunction with vanadium oxide, it is used as a catalyst for the oxidation of aromatic compounds in the synthesis of carboxylic acids and acid anhydrides.

Ceramic Glazes

Tin dioxide has long been used as an opacifier and as a white colourant in ceramic glazes. Its use has been particularly common in glazes for earthenware, sanitaryware and wall tiles; see the articles tin-glazing and Tin-glazed pottery. Tin oxide remains in suspension in vitreous matrix of the fired glazes, and, with its high refractive index being sufficiently different from the matrix, light is scattered, and hence increases the opacity of the glaze. The degree of dissolution increases with the firing temperature, and hence the extent of opacity diminishes. Although dependent on the other constituents the solubility of tin oxide in glaze melts is generally low. Its solubility is increased by Na_2O, K_2O and B_2O_3, and reduced by CaO, BaO, ZnO, Al_2O_3, and to a limited extent PbO.

SnO_2 has been used as pigment in the manufacture of glasses, enamels and ceramic glazes. Pure SnO_2 gives a milky white colour; other colours are achieved when mixed with other metallic oxides e.g. V_2O_5 yellow; Cr_2O_3 pink; and Sb_2O_5 grey blue.

Polishing

Tin dioxide can be used as a polishing powder, sometimes in mixtures also with lead oxide, for polishing glass, jewelery, marble and silver. Tin dioxide for this use is sometimes called as "putty powder" or "jeweler's putty".

Glass Coatings

SnO_2 coatings can be applied using chemical vapor deposition, vapour deposition techniques that employ $SnCl_4$ or organotin trihalides e.g. butyltin trichloride as the volatile agent. This technique is used to coat glass bottles with a thin (<0.1 μm) layer of SnO_2, which helps to adhere a subsequent, protective polymer coating such as polyethylene to the glass.

Thicker layers doped with Sb or F ions are electrically conducting and used in electroluminescent devices.

Gas Sensing

SnO_2 wires are commonly used as the detecting element in carbon monoxide detectors.

SnO_2 is used in sensors of combustible gases. In these the sensor area is heated to a constant temperature (few hundred °C) and in the presence of a combustible gas the electrical resistivity drops. Doping with various compounds has been investigated (e.g. with CuO). Doping with cobalt and manganese, gives a material that can be used in e.g. high voltage varistors. Tin dioxide can be doped into the oxides of iron or manganese.

Cu_2O

Copper(I) oxide or cuprous oxide is the inorganic compound with the formula Cu_2O. It is one of the principal oxides of copper. This red-coloured solid is a component of some antifouling paints. The compound can appear either yellow or red, depending on the size of the particles. Copper(I) oxide is found as the reddish mineral cuprite.

Preparation

Copper(I) oxide may be produced by several methods. Most straightforwardly, it arises via the oxidation of copper metal:

$$4\ Cu + O_2 \rightarrow 2\ Cu_2O$$

Additives such as water and acids affect the rate of this process as well as the further oxidation to copper(II) oxides. It is also produced commercially by reduction of copper(II) solutions with sulfur dioxide. Aqueous cuprous chloride solutions react with base to give the same material. In all cases, the colour is highly sensitive to the procedural details.

Formation of copper(I) oxide is the basis of the Fehling's test and Benedict's test for reducing sugars. These sugars reduce an alkaline solution of a copper(II) salt, giving a bright red precipitate of Cu_2O. It forms on silver-plated copper parts exposed to moisture when the silver layer is porous or damaged. This kind of corrosion is known as red plague.

Little evidence exists for cuprous hydroxide, which is expected to rapidly undergo dehydration. A similar situation applies to the hydroxides of gold(I) and silver(I).

Copper(I) oxide can also be prepared by reacting a copper-ammonia complex with hydrogen peroxide.

Properties

The solid is diamagnetic. In terms of their coordination spheres, copper centres are 2-coordinated and the oxides are tetrahedral. The structure thus resembles in some sense the main polymorphs of SiO_2, and both structures feature interpenetrated lattices.

Copper(I) oxide dissolves in concentrated ammonia solution to form the colourless complex $[Cu(NH_3)_2]^+$, which is easily oxidized in air to the blue $[Cu(NH_3)_4(H_2O)_2]^{2+}$. It dissolves in hydrochloric acid to give solutions of $CuCl_2^-$. Dilute sulfuric acid and nitric acid produce copper(II) sulfate and copper(II) nitrate, respectively.

Cu_2O degrades to copper(II) oxide in moist air.

Structure

Cu_2O crystallizes in a cubic structure with a lattice constant a_l=4.2696 Å. The Cu atoms arrange in a fcc sublattice, the O atoms in a bcc sublattice. One sublattice is shifted by a quarter of the body diagonal. The space group is $Pn\overline{3}m$, which includes the point group with full octahedral symmetry.

Semiconducting Properties

In the history of semiconductor physics, Cu_2O is one of the most studied materials, and many experimental observations and semiconductor applications have been demonstrated first in this material:

- Semiconductor
- Semiconductor diodes
- Experimental demonstration of Wannier exciton series
- Polariton propagation beats in a solid
- Dynamical Stark effect of excitons
- Phonoritons

Today, it is still under investigation in semiconductor optics. Particularly people are trying to create a Bose–Einstein condensate of excitons in Cu_2O.

The lowest excitons in Cu_2O are extremely long living and show a neV resonance, which is the narrowest bulk exciton resonance ever observed. The associated quadrupole polaritons have low group velocity approaching the speed of sound. Thus, light moves almost as slowly as sound in this medium, which results in high polariton densities.

Another unusual feature of the ground state excitons is that all primary scattering mechanisms are known quantitatively.

Cu_2O was the first substance where an entirely parameter-free model of absorption linewidth broadening by temperature could be established, allowing the corresponding absorption coefficient to be deduced. It can be shown using Cu_2O that the Kramers–Kronig relations do not apply to polaritons.

Applications

Cuprous oxide is commonly used as a pigment, a fungicide, and an antifouling agent for marine paints. Rectifier diodes based on this material have been used industrially as early as 1924, long before silicon became the standard. Copper(I) oxide is also responsible for the pink colour in a positive Benedict's Test.

Ga_2O_3

Gallium(III) oxide (Ga_2O_3) is a chemical compound used in vacuum deposition and as part of the manufacturing of semiconductor devices.

Preparation

Gallium oxide is precipitated in hydrated form upon neutralization of acidic or basic solution of gallium salt. Also, it is formed on heating gallium in air or by thermally decomposing gallium nitrate at 200-250ÚC. It can occur in five different modifications, α,β,δ,γ and ε. Of these modifications β-Ga_2O_3 is the most stable form.

Preparation Methods for the Five Modifications

- β-Ga_2O_3 is prepared by heating nitrate, acetate, oxalate or other organic derivatives above 1000°C.
- α-Ga_2O_3 can be obtained by heating β-Ga_2O_3 at 65kbars and 1100°C for 1 hour giving a crystalline structure. The hydrated form can be prepared by decomposing precipitated and "aged" gallium hydroxide at 500°C.
- γ-Ga_2O_3 is prepared by rapidly heating the hydroxide gel at 400°C-500ÚC. A more crystalline form of this polymorph can be prepared directly from gallium metal by a solvothermal synthesis.
- δ-Ga_2O_3 is obtained by heating $Ga(NO_3)_3$ at 250°C.
- ε-Ga_2O_3 is prepared by briefly heating δ-Ga_2O_3 at 550°C for 30 minutes.

Reactions

Gallium(III) oxide is amphoteric. reacting with alkali metal oxides at high temperature to form e.g. $NaGaO_2$, and with Mg, Zn, Co, Ni, Cu oxides to form spinels e.g. $MgGa_2O_4$. and dissolving in strong alkali to form a solution of the gallate ion, $Ga(OH)_{-4.}$

With HCl gas under argon it forms gallium trichloride $GaCl_3$. $Ga_2O_3 + 6\ HCl \rightarrow GaCl_3 + H_2O$

It can be reduced to gallium suboxide, (gallium(I) oxide), Ga_2O by H_2, or by reaction with gallium metal

$$Ga_2O_3 + H_2 \rightarrow Ga_2O + H_2O$$

$$Ga_2O_3 + 4\ Ga \rightarrow 3\ Ga_2O$$

Crystal Structure

β-Ga_2O_3, with a melting point of 1740°C, is the most stable crystalline modification. The oxide ions are in a distorted cubic closest packing arrangement, and the gallium (III) ions are in distorted tetrahedral and octahedral sites, with Ga-O bond distances of 1.83 and 2.00 A respectively. These distortions are in fact the reasons for the great level of stability of β-Ga_2O_3.

α-Ga_2O_3 has the same structure (corundum) as α-Al_2O_3 where all Ga atoms are 6 coordinate. γ-Ga_2O_3 has a defect spinel structure similar to that of γ-Al_2O_3.

Applications and Uses

Gallium(III) oxide is an important functional material. It has been studied in the use of lasers, phosphors and luminescent materials, has been shown to demonstrate catalytic properties and has also been used as an insulating barrier in tight junctions. The stable oxide of gallium, monoclinic β-Ga_2O_3, has current applications in gas sensors and luminescent phosphors and can be applied to dielectric coatings for solar cells. This stable oxide has also shown potential for deep-ultraviolet transparent conductive oxides.

Nanotechnology

Nanoribbons and nanosheets of Ga_2O_3 can be synthesized either by high temperature reaction of Ga^0 with water or by evaporation of GaN at high temperature in the presence of oxygen. Analysis of the products of the thermal evaporation reaction is done using X-ray diffraction, scanning electron microscopy (SEM), transmission electron microscopy (TEM) and energy-dispersive X-ray spectroscopy (EDS). Such analysis showed that the resulting structures were wool-like.

SEM reveals that the products consist of wire-like nanostructures and sheet-type structures. The TEM image shows the ribbon-like structure of Ga_2O_3. The EDS confirms that the nanostructures obtained are in fact Ga_2O_3.

The nanoribbon and nanosheets structures of Ga_2O_3 are pure, structurally uniform, single crystalline and free from dislocation. The structure of nanoribbons and nanosheets, that is, their wave-like and sheet-like shape, also indicates that their growth is as a result of growth kinetics, vapor-liquid-solid (VLS) method and vapor-solid method (VS). VLS and VS are two common growth mechanisms for nanowires.

The VLS process, catalytic-assisted in nature, is one in which the metal particle is located at the growth of the wire and acts as the catalytic active site. For the VS process, oxide vapor, which is evaporated from the starting oxide at a higher temperature xone, directly deposits on a substrate at a lower temperature region and grows into ribbon-like nanostructures.

Optical Uses

It is important to accurately determine the optical functions as these are essential for device simulations and improvement in material preparation. The thin Ga_2O_3 films are of commercial interest as gas sensitive material and Ga_2O_3 based glasses are among the best optical materials used in advanced technologies. Ellipsometry is a procedure that can be used to determine optical functions of the β-Ga_2O_3.

Catalyst

β-Gallium(III) oxide is also very important in the production of catalysts. It is needed for the preparation of Ga_2O_3-Al_2O_3 catalyst. The preparation of this catalyst involves reacting Al_2O_3 with aqueous solutions of gallium nitrate, followed by evaporation to dryness at 393K, and calcining in air (i.e.thermal decomposition of the compound) for 4 hours at 823K.

Fe_2O_3

Iron(III) oxide or ferric oxide is the inorganic compound with the formula Fe_2O_3. It is one of the three main oxides of iron, the other two being iron(II) oxide (FeO), which is rare, and iron(II,III) oxide (Fe_3O_4), which also occurs naturally as the mineral magnetite. As the mineral known as hematite, Fe_2O_3 is the main source of the iron for the steel industry. Fe_2O_3 is ferromagnetic, dark red, and readily attacked by acids. Iron(III) oxide is often called rust, and to some extent this label is useful, because rust shares several properties and has a similar composition. To a chemist, rust is considered an ill-defined material, described as *hydrated* ferric oxide.

Structure

Fe_2O_3 can be obtained in various polymorphs. In the main ones, α and γ, iron adopts octahedral coordination geometry. That is, each Fe centre is bound to six oxygen ligands.

Alpha Phase

α-Fe_2O_3 has the rhombohedral, corundum (α-Al_2O_3) structure and is the most common form. It occurs naturally as the mineral hematite which is mined as the main ore of iron. It is antiferromagnetic below

~260 K (Morin transition temperature), and exhibits weak ferromagnetism between 260 K and the Néel temperature, 950 K. It is easy to prepare using both thermal decomposition and precipitation in the liquid phase. Its magnetic properties are dependent on many factors, e.g. pressure, particle size, and magnetic field intensity.

Gamma Phase

γ-Fe_2O_3 has a cubic structure. It is metastable and converted from the alpha phase at high temperatures. It occurs naturally as the mineral maghemite. It is ferromagnetic and finds application in recording tapes, although ultrafine particles smaller than 10 nanometres are superparamagnetic. It can be prepared by thermal dehydratation of gamma iron(III) oxide-hydroxide, careful oxidation of iron(II,III) oxide. Another method involves the careful oxidation of Fe_3O_4. The ultrafine particles can be prepared by thermal decomposition of iron(III) oxalate.

Other Phases

Several other phases have been identified or claimed. The beta-phase is cubic body centred (space group Ia3), metastable, and at temperatures above 500 °C (930 °F) converts to alpha phase. It can be prepared by reduction of hematite by carbon, pyrolysis of iron(III) chloride solution, or thermal decomposition of iron(III) sulfate. The epsilon phase is rhombic, and shows properties intermediate between alpha and gamma. So far has not been prepared in pure form; it is always mixed with the alpha phase or gamma phases. Material with a high proportion of epsilon phase can be prepared by thermal transformation of the gamma phase. This phase is also metastable, transforming to the alpha phase at between 500 and 750 °C (930 and 1,380 °F). Can also be prepared by oxidation of iron in an electric arc or by sol-gel precipitation from iron(III) nitrate. Additionally at high pressure an amorphous form is claimed.

Hydrated iron(III) Oxides

Several hydrates of Iron(III) oxide exists. When alkali is added to solutions of soluble Fe(III) salts, a red-brown gelatinous precipitate forms. This is *not* $Fe(OH)_3$, but $Fe_2O_3{\cdot}H_2O$ (also written as Fe(O)OH). Several forms of the hydrated oxide of Fe(III) exist as well. The red *lepidocrocite* γ-Fe(O)OH, occurs on the outside of rusticles, and the orange *goethite*, which occurs internally in rusticles. When $Fe_2O_3{\cdot}H_2O$ is heated, it loses its water of hydration. Further heating at 1670 K converts Fe_2O_3 to black Fe_3O_4 ($Fe^{II}Fe^{III}{}_2O_4$), which is known as the mineral *magnetite*. Fe(O)OH is soluble in acids, giving $[Fe(OH_2)_6]^{3+}$. In concentrated aqueous alkali, Fe_2O_3 gives $[Fe(OH)_6]^{3-}$.

Reactions

The most important reaction is its carbothermal reduction, which gives iron used in steel-making:

$$2\ Fe_2O_3 + 3\ C \rightarrow 4\ Fe + 3\ CO_2$$

Another redox reaction is the extremely exothermic thermite reaction with aluminium.

$$2\ Al + Fe_2O_3 \rightarrow 2\ Fe + Al_2O_3$$

This process is used to weld thick metals such as rails of train tracks by using a ceramic container to funnel the molten iron in between two sections of rail. Thermite is also used in weapons and making small-scale cast-iron sculptures and tools.

Partial reduction with hydrogen at about 400 °C gives magnetite, a black magnetic material that contains both Fe(III) and Fe(II):

$$3\ Fe_2O_3 + H_2 \rightarrow 2\ Fe_3O_4 + H_2O$$

Iron(III) oxide is insoluble in water [needs citatation] but dissolves readily in strong acid, e.g. hydrochloric and sulfuric acids. It also dissolves well in solutions of the chelating agents such as EDTA and oxalic acid. Heating iron(III) oxides with other metal oxides or carbonates yields materials known as ferrates:

$$ZnO + Fe_2O_3 \rightarrow Zn(FeO_2)_2$$

Preparation

Iron (III) oxide is a product of the oxidation of iron. It can be prepared in the laboratory by electrolyzing a solution of sodium bicarbonate, an inert electrolyte, with an iron anode:

$$4\ Fe + 3\ O_2 + 2\ H_2O \rightarrow 4\ FeO(OH)$$

The resulting hydrated iron(III) oxide, written here as Fe(O)OH, dehydrates around 200 °C.

$$2\ FeO(OH) \rightarrow Fe_2O_3 + H_2O$$

It can also be prepared by the thermal decomposition of iron (III) hydroxide under temperature above 200 °C.

$$2\ Fe(OH)_3 \rightarrow Fe_2O_3 + 3H_2O$$

Uses

Iron Industry

The overwhelming application of iron(III) oxide is as the feedstock of the steel and iron industries, e.g. the production of iron, steel, and many alloys.

Polishing

A very fine powder of ferric oxide is known as "jeweler's rouge", "red rouge", or simply rouge. It is used to put the final polish on metallic jewellery and lenses, and historically as a cosmetic. Rouge cuts more slowly than some modern polishes, such as cerium(IV) oxide, but is still used in optics fabrication and by jewelers for the superior finish it can produce. When polishing gold, the rouge slightly stains the gold, which contributes to the appearance of the finished piece. Rouge is sold as a powder, paste, laced on polishing cloths, or solid bar (with a wax or grease binder). Other polishing compounds are also often called "rouge", even when they do not contain iron oxide. Jewelers remove the residual rouge on jewellery by use of ultrasonic cleaning. Products sold as "stropping compound" are often applied to a leather strop to assist in getting a razor edge on knives, straight razors, or any other edged tool.

Pigment

Iron(III) oxide is also used as a pigment, under names "Pigment Brown 6", "Pigment Brown 7", and "Pigment Red 101". Some of them, e.g. Pigment Red 101 and Pigment Brown 6, are Food and Drug Administration (FDA)-approved for use in cosmetics. Iron oxides are used as pigments in dental composites alongside titanium oxides.

Hematite is the characteristic component of the Swedish paint colour Falu red.

Magnetic Recording

Iron(III) oxide was the most common magnetic particle used in all types of magnetic storage and recording media, including magnetic disks (for data storage) and magnetic tape (used in audio and video recording as well as data storage). However, modern magnetic storage media - in particular, the hard disk drives - use more advanced thin film technology, which may consist of a stack of 15 layers or more.

Photocatalyst

α-Fe_2O_3 has been studied as a photoanode for the water-splitting reaction for over 25 years.

In_2O_3

Indium(III) oxide (In_2O_3) is a chemical compound, an amphoteric oxide of indium.

Physical Properties

Crystal Structure: Amorphous indium oxide is insoluble in water but soluble in acids, whereas crystalline indium oxide is insoluble in both water and acids. The crystalline form exist in two phases, the cubic (bixbyite type) and rhombohedral (corundum type). Both phases have a band gap of about 3 eV. The parameters of the cubic phase are listed in the infobox. The rhombohedral phase is produced at high temperatures and pressures or when using non-equilibrium growth methods. It has a space group R3c No. 167, Pearson symbol hR30, a = 0.5487 nm, b = 0.5487 nm, c = 0.57818 nm, Z = 6 and calculated density 7.31 g/cm^3.

Conductivity and Magnetism

Thin films of chromium-doped indium oxide ($In_{2-x}Cr_xO_3$) are a magnetic semiconductor displaying high-temperature ferromagnetism, single-phase crystal structure, and semiconductor behaviour with high concentration of charge carriers. It has possible applications in spintronics as a material for spin injectors.

Thin polycrystalline films of indium oxide doped with Zn are highly conductive (conductivity ~10^5 S/m) and even superconductive at helium temperatures. The superconducting transition temperature Tc depends on the doping and film structure and is below 3.3 K.

Synthesis

Bulk samples can be prepared by heating indium(III) hydroxide or the nitrate, carbonate or sulfate. Thin films of indium oxide can be prepared by sputtering of indium target in argon/oxygen atmosphere. They can be used as diffusion barriers ("barrier metals") in semiconductors, e.g. to inhibit diffusion between aluminium and silicon.

Monocrystalline nanowires were synthetized from indium oxide by laser ablation, allowing precise diameter control down to 10 nm. Field effect transistors were fabricated from those. Indium oxide nanowires can serve as sensitive and specific redox protein sensors. Sol-gel method is another way to prepare the nanowires. Indium oxide can serve as a semiconductor material, forming heterojunctions with p-InP, n-GaAs, n-Si, and other materials.

A layer of indium oxide on a silicon substrate can be deposited from an indium trichloride solution, a method useful for manufacture of solar cells.

Reactions

When heated to 700°C Indium(III) oxide forms In_2O, (called indium(I) oxide or indium suboxide), at 2000°C it decomposes. It is soluble in acids but not in alkali. With ammonia at high temperature indium nitride is formed

$$In_2O_3 + 2NH_3 \rightarrow 2InN + 3H_2O$$

With K_2O and indium metal the compound K_5InO_4 containing tetrahedral InO_4^{5-} ions was prepared. Reacting with a range of metal trioxides produced perovskites for example:- $In_2O_3 + Cr_2O_3 \rightarrow 2InCrO_3$

Applications

Indium oxide is used in some types of batteries, thin film infrared reflectors transparent for visible light (hot mirrors), some optical coatings, and some antistatic coatings. In combination with tin dioxide, indium oxide forms indium tin oxide (also called tin doped indium oxide or ITO), a material used for transparent conductive coatings.

In semiconductors, indium oxide can be used as an n-type semiconductor used as a resistive element in integrated circuits. In histology, indium oxide is used as a part of some stain formulations.

CdO

Cadmium oxide is an inorganic compound with the formula CdO. It is one of the main precursors to other cadmium compounds. It crystallizes in a cubic rocksalt lattice like sodium chloride, with octahedral cation and anion centres. It occurs naturally as the rare mineral monteponite. Cadmium oxide can be found as a colourless amorphous powder or as brown or red crystals. Cadmium oxide is an n-type semiconductor with a band gap of 2.16 eV at room temperature.

Production and Structure

Since cadmium compounds are often found in association with zinc ores, cadmium oxide is a common by-product of zinc refining. It is produced by burning elemental cadmium in air. Pyrolysis of other cadmium compounds, such as the nitrate or the carbonate, also affords this oxide. When pure, it is red but CdO is unusual in being available in many differing colours due to its tendency to form defect structures resulting from anion vacancies. Cadmium oxide is prepared commercially by oxidizing cadmium vapor in air.

Uses

Cadmium oxide is used in cadmium plating baths, electrodes for storage batteries, cadmium salts, catalyst, ceramic glazes, phosphors, and nematocide. Major uses for cadmium oxide are as an ingredient for electroplating baths, and in pigments.

Transparent Conductor

CdO is used as a transparent conductive material, which was prepared as a transparent conducting film back in 1907. Cadmium oxide in the form of thin films has been used in applications such as photodiodes, phototransistors, photovoltaic cells, transparent electrodes, liquid crystal displays, IR detectors, and anti reflection coatings. CdO microparticles undergo bandgap excitation when exposed to UV-A light and is also selective in phenol photodegradation.

Cadmium Plating

Most commercial electroplating of cadmium is done by electrodeposition from cyanide baths. These cyanide baths consist of cadmium oxide and sodium cyanide in water, which likely form cadmium cyanide and sodium hydroxide.

A typical formula is 32 g/L cadmium oxide and 75 g/L sodium cyanide. The cadmium concentration may vary by as much as 50%. Brighteners are usually added to the bath and the plating is done at room temperature with high purity cadmium anodes.

Reactivity

CdO is a basic oxide and is thus attacked by aqueous acids to give solutions of $[Cd(H_2O)_6]^{2+}$. Upon treatment with strong alkaline solutions, [Cd(OH)4]2– forms. A thin coat of cadmium oxide forms on the surface of cadmium in moist air at room temperature. Cadmium will oxidize at room temperatures to form CdO. Cadmium vapor and steam will form CdO and hydrogen in a reversible reaction.

CeO_2

Cerium(IV) oxide, also known as ceric oxide, ceria, cerium oxide or cerium dioxide, is an oxide of the rare earth metal cerium. It is a pale yellow-white powder with the chemical formula CeO_2.

Cerium(IV) oxide is formed by the calcination of cerium oxalate or cerium hydroxide.

Powdered ceria is slightly hygroscopic and will also absorb a small amount of carbon dioxide from the atmosphere.

Cerium also forms cerium(III) oxide, Ce2O3, which is more stable at standard conditions for temperature and pressure than CeO2.

Structure

Cerium oxide has the fluorite structure, space group Fm3m, #225 containing 8 coordinate Ce^{4+} and 4 coordinate O^{2-}. At high temperatures it can be reduced to an non-stoichiometric, anion deficient form that retains the fluorite lattice, $CeO_{(2-x)}$ where $0 < x < 0.28$ The non stoichiometric form has a blue to black colour, and exhibits both ionic and electronic conduction with ionic being the most significant at temperatures > 500 °C.

Applications

Cerium(IV) oxide is used in ceramics, to sensitize photosensitive glass, as a catalyst and as a catalyst support, to polish glass and stones, in lapidary as an alternative to "jeweller's rouge". It is also known as "optician's rouge".

It is also used in the walls of self-cleaning ovens as a hydrocarbon catalyst during the high-temperature cleaning process.

While it is transparent for visible light, it absorbs ultraviolet radiation strongly, so it is a prospective replacement of zinc oxide and titanium dioxide in sunscreens, as it has lower photocatalytic activity. However, its thermal catalytic properties have to be decreased by coating the particles with amorphous silica or boron nitride. The use of these nanoparticles, which can penetrate the body and reach internal organs, has been criticized as unsafe.

Cerium oxide has found use in infrared filters, as an oxidizing species in catalytic converters and as a replacement for thorium dioxide in incandescent mantles.

Fuel Cell Electrolyte

In the doped form (it comes from cerium and oxygen), ceria is of interest as a material for solid oxide fuel cells or SOFCs because of its relatively high oxygen ion conductivity (i.e. oxygen atoms readily move through it) at intermediate temperatures (500–650 °C). Undoped and doped ceria also exhibit high electronic conductivity at low partial pressures of oxygen due to reduction of the cerium ion leading to the formation of small polarons. However, doped ceria has an extended electrolytic region (area of predominant ionic conductivity), over that of ceria, that allows its use as an electrolyte in SOFCs below 600-650 °C. Exposure to hydrogen at high (800 °C) has been shown to cause significant damage to the grain boundaries leading to cracking. Exposure to other reducing agents such as carbon monoxide is less damaging. Substituting a fraction of the ceria with gadolinium (as in Gadolinium doped ceria) or samarium will introduce oxygen vacancies in the crystal without adding electronic charge carriers. This increases the ionic conductivity and results in a better electrolyte.

Under reducing conditions, such as those experienced on the anode side of the fuel cell, a large amount of oxygen vacancies within the ceria electrolyte can be formed. Some of the cerium(IV) oxide is also reduced to cerium(III) oxide under these conditions, which consequently increases the electronic conductivity of the material. The lattice constant of ceria increases under reducing conditions as well as with decreasing nanocrystal size in nanocrystalline ceria, as a result of reduction of the cerium cation from a 4+ to a 3+ state in order to charge compensate for oxygen vacancy formation.

Catalyst

Ceria has been used in catalytic converters in automotive applications. Since ceria can become non-stoichiometric in oxygen content (i.e. it can give up oxygen without decomposing) depending on its ambient partial pressure of oxygen, it can release or take in oxygen in the exhaust stream of a combustion engine. In association with other catalysts, ceria can effectively reduce NO_x emissions as well as convert harmful carbon monoxide to the less harmful carbon dioxide. Ceria is particularly interesting for catalytic conversion economically because it has been shown that adding comparatively inexpensive ceria can allow for substantial reductions in the amount of platinum needed for complete oxidation of NO_x and other harmful products of incomplete combustion.

Due to its fluorite structure, the oxygen atoms in a ceria crystal are all in a plane with one another, allowing for rapid diffusion as a function of the number of oxygen vacancies. As the number of vacancies increases, the ease at which oxygen can move around in the crystal increases, allowing the ceria to reduce and oxidize molecules or co-catalysts on its surface. It has been shown that the catalytic activity of ceria is directly related to the number of oxygen vacancies in the crystal, frequently measured by using X-ray photoelectron spectroscopy to compare the ratios of Ce3+to Ce4+in the crystal.

Ceria can also be used as a co-catalyst in a number of reactions, including the water-gas shift reaction and steam reforming of ethanol or diesel fuel into hydrogen gas and carbon dioxide (with varying combinations of rhodium oxide, iron oxide, cobalt oxide, nickel oxide, platinum, and gold), the Fischer-Tropsch reaction, and selected oxidation (particularly with lanthanum). In each case, it has been shown that increasing the ceria oxygen defect concentration will result in increased catalytic activity, making it very interesting as a nanocrystalline co-catalyst due to the heightened number of oxygen defects as crystallite size decreases—at very small sizes, as many as 10% of the oxygen sites in the fluorite structure crystallites will be vacancies, resulting in exceptionally high diffusion rates.

Water Splitting

The cerium(IV) oxide–cerium(III) oxide cycle or CeO_2/Ce_2O_3 cycle is a two step thermochemical water splitting process based on cerium(IV) oxide and cerium(III) oxide for hydrogen production.

Anti-Oxidant

Uniform 3.8-nanometre spheres of cerium oxide with a thin coating of fatty oleic acid make them biocompatible. The spheres act as anti-oxidants, absorbing ROS free radicals. One gram of these nanoparticles can have the surface area of a football field. Potential uses include treatments for traumatic brain injury, cardiac arrest, Alzheimer's disease and could help to reduce palliating radiation-induced side effects suffered by cancer patients. The nanoparticles also have potential to protect astronauts from long-term exposure to radiation in space and perhaps even slow the effects of aging.

The particles continue to work over time by reverting to their initial state, for unexplained reasons.

The coating is thin enough to let oxygen pass through to the particle, but robust enough to protect it through many cycles of ROS absorption. The particles are uniform with well-defined surfaces. They are made without water, which maximizes the surface gaps available for oxygen scavenging.

Defects

In the most stable fluorite phase of ceria, it exhibits several defects depending on partial pressure of oxygen or stress state of the material. The primary defects of concern are oxygen vacancies and small polarons (electrons localized on cerium cations) because these two are located in the "useful" range of ceria. In the case of oxygen defects, the increased diffusion rate of oxygen in the lattice causes increased catalytic activity as well as an increase in ionic conductivity, making ceria interesting as a fuel cell electrolyte in solid-oxide fuel cells.

3

Surface Plasmon of Metal Nanoparticles: Theory and Sensing Applications

Surface plasmons (SPs) are coherent electron oscillations that exist at the interface between any two materials where the real part of the dielectric function changes sign across the interface (e.g. a metal-dielectric interface, such as a metal sheet in air). SPs have lower energy than bulk (or volume) plasmons which quantise the longitudinal electron oscillations about positive ion cores within the bulk of an electron gas (or plasma).

The charge motion in a surface plasmon always creates electromagnetic fields outside (as well as inside) the metal. The *total* excitation, including both the charge motion and associated electromagnetic field, is called either a surface plasmon polariton at a planar interface, or a localized surface plasmon for the closed surface of a small particle.

The existence of surface plasmons was first predicted in 1957 by Rufus Ritchie. In the following two decades, surface plasmons were extensively studied by many scientists, the foremost of whom were T. Turbadar in the 1950s and 1960s, and Heinz Raether, E. Kretschmann, and A. Otto in the 1960s and 1970s. Information transfer in nanoscale structures, similar to photonics, by means of surface plasmons, is referred to as plasmonics.

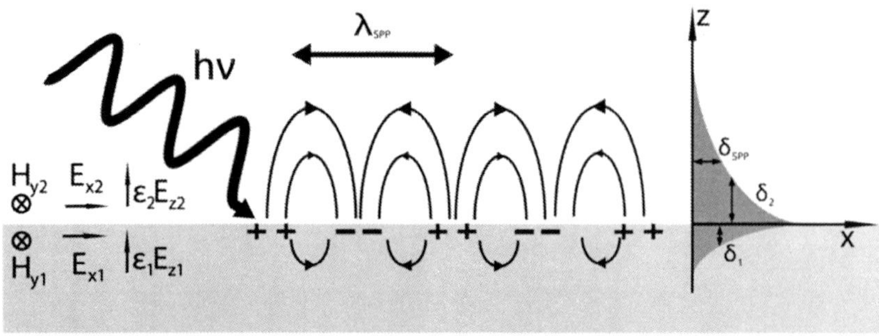

Figure: *Schematic representation of an electron density wave propagating along a metal – dielectric interface. The charge density oscillations and associated electromagnetic fields are called surface plasmon-polariton waves. The exponential dependence of the electromagnetic field intensity on the distance away from the interface is shown on the right. These waves can be excited very efficiently with light in the visible range of the electromagnetic spectrum.*

Surface Plasmon Polaritons

Surface plasmon polaritons (SPPs), are infrared or visible-frequency electromagnetic waves, which travel along a metal-dielectric or metal-air interface. The term "surface plasmon polariton" explains that the wave involves both charge motion in the metal ("surface plasmon") and electromagnetic waves in the air or dielectric ("polariton").

They are a type of surface wave, guided along the interface in much the same way that light can be guided by an optical fibre. SPPs are shorter in wavelength than the incident light (photons). Hence, SPPs can have tighter spatial confinement and higher local field intensity. Perpendicular to the interface, they have subwavelength-scale confinement. An SPP will propagate along the interface until its energy is lost either to either absorption in the metal or scattering into other directions (such as into free space).

Application of SPPs enables subwavelength optics in microscopy and lithography beyond the diffraction limit. It also enables the first steady-state micro-mechanical measurement of a fundamental property of light itself: the momentum of a photon in a dielectric medium. Other applications are photonic data storage, light generation, and bio-photonics.

Excitation

SPPs can be excited by both electrons and photons. Excitation by electrons is created by firing electrons into the bulk of a metal. As the electrons scatter, energy is transferred into the bulk plasma. The component of the scattering vector parallel to the surface results in the formation of a surface plasmon polariton.

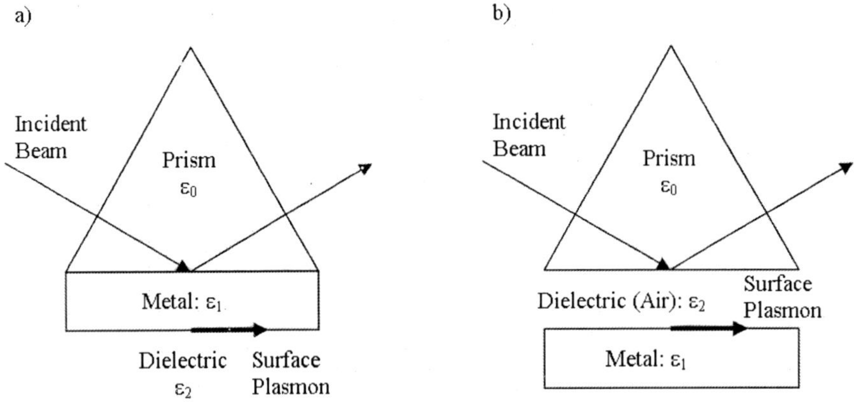

***Figure:** (a) Kretschmann and (b) Otto configuration of an Attenuated Total Reflection setup for coupling surface plasmons. In both cases, the surface plasmon propagates along the metal/dielectric interface*

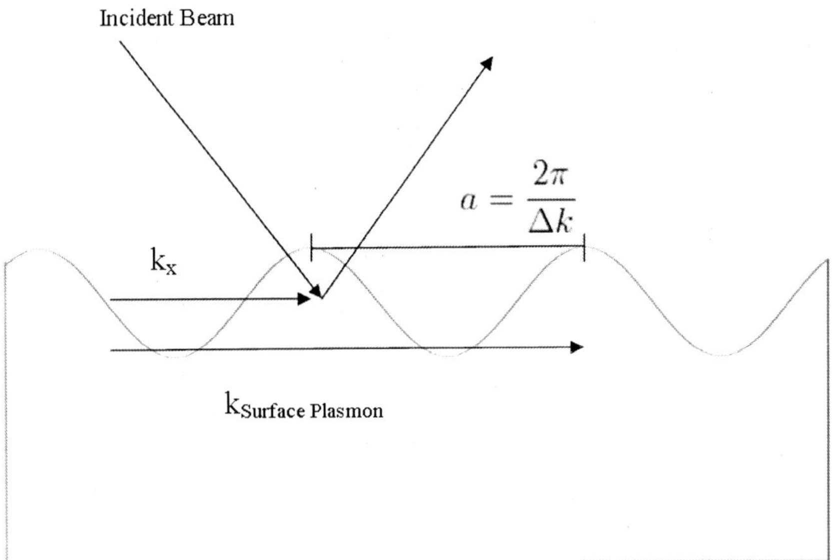

***Figure:** Grating Coupler for Surface Plasmons. The wave vector is increased by the spatial frequency*

If a free-space photon comes from air towards a smooth metal surface, it *cannot* excite an SPP at the metal-air interface. The reason is that if the photon and SPP have the same frequency, then they necessarily have different in-plane wavevectors. This incompatibility is analogous to the lack of transmission that occurs during total internal reflection. Analogously, an SPP on a smooth metal surface *cannot* lose energy to radiation into the dielectric (if the dielectric is uniform).

Nevertheless, coupling of photons into SPPs can be achieved using a coupling medium such as a prism or grating to match the photon and surface plasmon wave vectors. A prism can be positioned against a thin metal film in the Kretschmann configuration or very close to a metal surface in the Otto configuration. A grating coupler matches the wave vectors by increasing the parallel wave vector component by an amount related to the grating period. This method, while less frequently utilised, is critical to the theoretical understanding of the effect of surface roughness. Moreover, simple isolated surface defects such as a groove, a slit or a corrugation on an otherwise planar surface provides a mechanism by which free-space radiation and SPs can exchange energy and hence couple.

Dispersion Relation

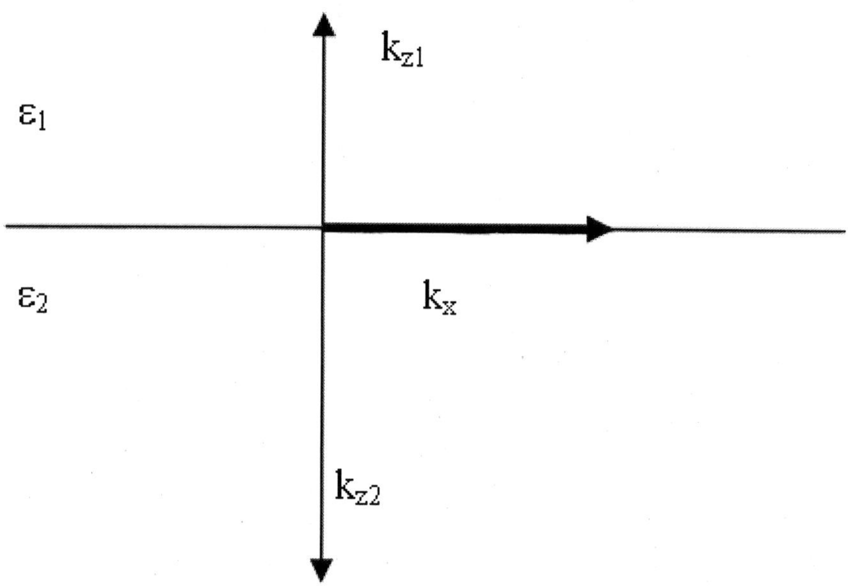

Figure: *Coordinate system for 2 material interface*

The electric field of a propagating electromagnetic wave can be expressed

$$E = E_0 \exp[i(k_x x + k_z z - \omega t)]$$

where k is the wave number and ω is the frequency of the wave. By solving Maxwell's equations for the electromagnetic wave at an interface between two materials with relative dielectric functions ε_1 and ε_2 with the appropriate continuity relation the boundary conditions are

$$\frac{k_{z1}}{\varepsilon_1} + \frac{k_{z2}}{\varepsilon_2} = 0$$

and

$$k_x^2 + k_{zi}^2 = \varepsilon_i \left(\frac{\omega}{c}\right)^2 \qquad i = 1,2$$

where c is the speed of light in a vacuum, and k_x is same for both media at the interface for a surface wave. Solving these two equations, the dispersion relation for a wave propagating on the surface is

$$k_x = \frac{\omega}{c}\left(\frac{\varepsilon_1 \varepsilon_2}{\varepsilon_1 + \varepsilon_2}\right)^{1/2}.$$

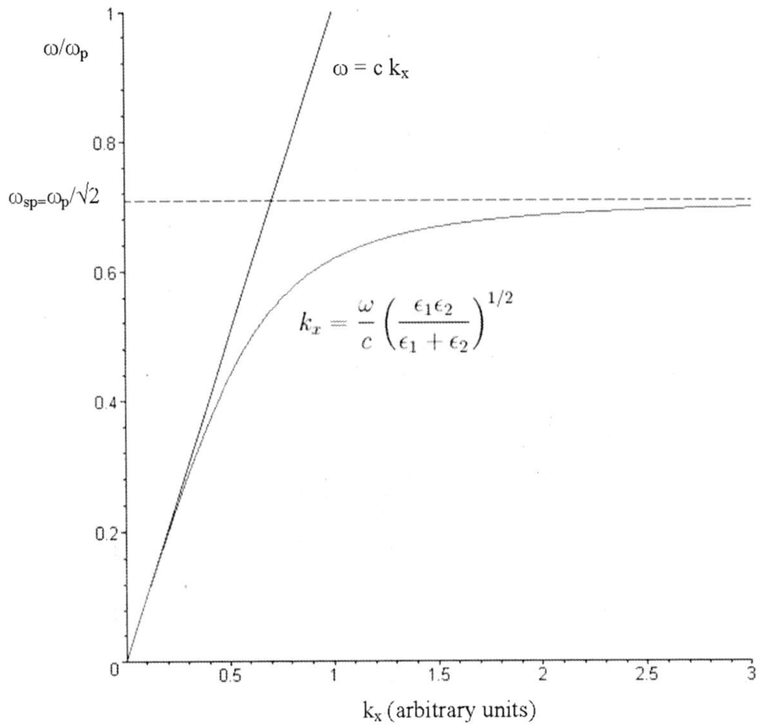

Figure: *Dispersion curve for surface plasmon polaritons. At low* k, *the surface plasmon curve (red) approaches the photon curve (blue)*

In the free electron model of an electron gas, which neglects attenuation, the metallic dielectric function is

$$\varepsilon(\omega) = 1 - \frac{\omega_P^2}{\omega^2},$$

where the bulk plasma frequency in SI units is

$$\omega_P = \sqrt{\frac{ne^2}{\varepsilon_0 m^*}}$$

where n is the electron density, e is the charge of the electron, m^* is the effective mass of the electron and ε_0 is the permittivity of free-space. The dispersion relation is plotted in Figure 4. At low k, the SPP behaves like a photon, but as k increases, the dispersion relation bends over and reaches an asymptotic limit called the "surface plasma frequency". Since the dispersion curve lies to the right of the light line, $\omega = k \cdot c$, the SPP has a shorter wavelength than free-space radiation such that the out-of-plane component of the SPP wavevector is purely imaginary and exhibits evanescent decay. The surface plasma frequency is the asymptote of this curve, and is given by

$$\omega_{SP} = \omega_P / \sqrt{1+\varepsilon_2}.$$

In the case of air, this result simplifies to

$$\omega_{SP} = \omega_P / \sqrt{2}.$$

If we assume that ε_2 is real and $\varepsilon_2 > 0$, then it must be true that $\varepsilon_1 < 0$, a condition which is satisfied in metals. Electromagnetic waves passing through a metal experience damping due to Ohmic losses and electron-core interactions. These effects show up in as an imaginary component of the dielectric function. The dielectric function of a metal is expressed $\varepsilon_1 = \varepsilon_1' + i \cdot \varepsilon_1''$ where ε_1' and ε_1'' are the real and imaginary parts of the dielectric function, respectively. Generally $|\varepsilon_1'| >> \varepsilon_1''$ so the wavenumber can be expressed in terms of its real and imaginary components as

$$k_x = k_{x'} + ik_{x''} = \left[\frac{\omega}{c}\left(\frac{\varepsilon_{1'}\varepsilon_2}{\varepsilon_{1'}+\varepsilon_2}\right)^{1/2}\right] + i\left[\frac{\omega}{c}\left(\frac{\varepsilon_{1'}\varepsilon_2}{\varepsilon_{1'}+\varepsilon_2}\right)^{3/2}\frac{\varepsilon_{1''}}{2(\varepsilon_{1'})^2}\right].$$

The wave vector gives us insight into physically meaningful properties of the electromagnetic wave such as its spatial extent and coupling requirements for wave vector matching.

Propagation Length and Skin Depth

As an SPP propagates along the surface, it loses energy to the metal due to absorption. The intensity of the surface plasmon decays with the square of the electric field, so at a distance x, the intensity has decreased by a factor of $\exp[-2k_x''x]$. The propagation length is defined as the distance for the SPP intensity to decay by a factor of $1/e$. This condition is satisfied at a length

$$L = \frac{1}{2k_{x''}}.$$

Likewise, the electric field falls off evanescently perpendicular to the metal surface. At low frequencies, the SPP penetration depth into the metal is commonly approximated using the skin depth formula. In the dielectric, the field will fall off far more slowly. The decay lengths in the metal and dielectric medium can be expressed as

$$z_i = \frac{\lambda}{2\pi}\left(\frac{|\varepsilon_{1'}| + \varepsilon_2}{\varepsilon_i^2}\right)^{1/2}$$

where i indicates the medium of propagation. SPPs are very sensitive to slight perturbations within the skin depth and because of this, SPPs are often used to probe inhomogeneities of a surface.

Experimental Applications

Nanofabricated systems that exploit SPPs demonstrate potential for designing and controlling the propagation of light in matter. In particular, SPPs can be used to channel light efficiently into nanometre scale volumes, leading to direct modification of resonate frequency dispersion properties (substantially shrinking the wavelength of light and the speed of light pulses for example), as well as field enhancements suitable for enabling strong interactions with nonlinear materials. The resulting enhanced sensitivity of light to external parameters (for example, an applied electric field or the dielectric constant of an adsorbed molecular layer) shows great promise for applications in sensing and switching.

Current research is focused on the design, fabrication, and experimental characterization of novel components for measurement and communications based on nanoscale plasmonic effects. These devices include ultra-compact plasmonic interferometres for applications such as biosensing, optical positioning and optical switching, as well as the individual building blocks (plasmon source, waveguide and detector) needed to integrate a high-bandwidth, infrared-frequency plasmonic communications link on a silicon chip.

In addition to building functional devices based on SPPs, it appears feasible to exploit the dispersion characteristics of SPPs travelling in confined metallo-dielectric spaces to create photonic materials with artificially tailored bulk optical characteristics, otherwise known as *metamaterials*.

The excitation of SPPs is frequently used in an experimental technique known as surface plasmon resonance (SPR). In SPR, the maximum excitation of surface plasmons are detected by monitoring the reflected power from a prism coupler as a function of incident angle or wavelength. This technique can be used to observe nanometre changes in thickness, density fluctuations, or molecular absorption.

Surface plasmon-based circuits, including both SPPs and localized plasmon resonances, have been proposed as a means of overcoming the size limitations of photonic circuits for use in high performance data processing nano devices.

The ability to dynamically control the plasmonic properties of materials in these nano-devices is key to their development. A new approach that uses plasmon-plasmon interactions has been demonstrated recently. Here the bulk plasmon resonance is induced or suppressed to manipulate the propagation of light. This approach has been shown to have a high potential for nanoscale light manipulation and the development of a fully CMOS- compatible electro-optical plasmonic modulator.

CMOS compatible electro-optic plasmonic modulators will be key components in chip-scale photonic circuits.

In surface second harmonic generation, the second harmonic signal is proportional to the square of the electric field. The electric field is stronger at the interface because of the surface plasmon resulting in a non-linear optical effect. This larger signal is often exploited to produce a stronger second harmonic signal.

The wavelength and intensity of the plasmon-related absorption and emission peaks are affected by molecular adsorption that can be used in molecular sensors.

For example, a fully operational prototype device detecting casein in milk has been fabricated. The device is based on monitoring changes in plasmon-related absorption of light by a gold layer.

Effects of Roughness

In order to understand the effect of roughness on SPPs, it is beneficial to first understand how a SPP is coupled by a grating Figure earlier. When a photon is incident on a surface, the wave vector of the photon in the dielectric material is smaller than that of the SPP. In order for the photon to couple into a SPP, the wave vector must increase by $\Delta k = k_{SP} - k_{x,\text{photon}}$.

The grating harmonics of a periodic grating provide additional momentum parallel to the supporting interface to match the terms.

$$k_{SPP} = k_{x,\text{photon}} \pm n\, k_{\text{grating}} = \frac{\omega}{c}\sin\theta_0 \pm n\frac{2\pi}{a}$$

where k_{grating} is the wave vector of the grating, θ_0 is the angle of incidence of the incoming photon, a is the grating period, and n is an integer.

Rough surfaces can be thought of as the superposition of many gratings of different periodicities. Kretschmann proposed that a statistical correlation function be defined for a rough surface

$$G(x,y)=\frac{1}{A}\int_A z(x',y')\,z(x'-x,y'-y)dx'dy'$$

where $z(x,y)$ is the height above the mean surface height at the position (x,y), and A is the area of integration. Assuming that the statistical correlation function is Gaussian of the form

$$G(x,y)=\delta^2\exp\left(-\frac{r^2}{\sigma^2}\right)$$

where δ is the root mean square height, r is the distance from the point (x,y), and σ is the correlation length, then the Fourier transform of the correlation function is

$$|s(k_{\text{surf}})|^2=\frac{1}{4\pi}\sigma^2\delta^2\exp\left(-\frac{\sigma^2k_{\text{surf}}^2}{4}\right)$$

where s is a measure of the amount of each spatial frequency k_{surf} which help couple photons into a surface plasmon.

If the surface only has one Fourier component of roughness (i.e. the surface profile is sinusoidal), then the is discrete and exists only at $k=\frac{2\pi}{a}$, resulting in a single narrow set of angles for coupling. If the surface contains many Fourier components, then coupling becomes possible at multiple angles. For a random surface, becomes continuous and the range of coupling angles broadens.

As stated earlier, SPPs are non-radiative. When a SPP travels along a rough surface, it usually becomes radiative due to scattering. The Surface Scattering Theory of light suggests that the scattered intensity dI per solid angle $d\Omega$ per incident intensity I_0 is

$$\frac{dI}{d\Omega\, I_0}=\frac{4\sqrt{\varepsilon_0}}{\cos\theta_0}\frac{\pi^4}{\lambda^4}|t_{012}^p|^2\,|W|^2|s(k_{\text{surf}})|^2$$

where $|W|^2$ is the radiation pattern from a single dipole at the metal/dielectric interface. If surface plasmons are excited in the Kretschmann geometry and the scattered light is observed in the plane of incidence, then the dipole function becomes

$$|W|^2=A(\theta,|\varepsilon_1|)\sin^2\psi\,[(1+\sin^2\theta/|\varepsilon_1|)^{1/2}-\sin\theta]^2$$

with

$$A(\theta, |\varepsilon_1|) = \frac{|\varepsilon_1|+1}{|\varepsilon_1|-1} \frac{4}{1+\tan\theta/|\varepsilon_1|}$$

where ψ is the polarization angle and θ is the angle from the z-axis in the xz-plane. Two important consequences come out of these equations. The first is that if $\psi = 0$ (s-polarization), then $|W|^2 = 0$ and the scattered light $\frac{dI}{d\Omega\, I_0} = 0$. Secondly, the scattered light has a measurable profile which is readily correlated to the roughness. This topic is treated in greater detail in reference.

Metal Nanoparticles

Recent advances that allow metals to be structured and characterized on the nanometre scale have renewed the interest of physicists, chemists, materials scientists, and biologists in surface plasmons. Surface plasmons are collective excitations of the electrons at the interface between a conductor and an insulator and are described by evanescent electromagnetic waves that are not necessarily located at the interface. Surface plasmons appear in a number of different phenomena, including the optical response of materials at different scales, to Casimir and van der Waals forces between macroscopic bodies.

The strong coupling between light and surface plasmons in nanostructures leads to novel phenomena such as optical force enhancement in nanoaggregates, transport and storage of energy, as well as its localization and guiding, surface-enhanced sensing and spectroscopy, controlling the anisotropic growth of nanoparticles, and measuring intramolecular and conformational distances in molecules.

The understanding of surface plasmon resonances (SPRs) provides a design rule that guides the development of more complex nanostructures with an optimal optical response that we can both predict and experimentally realise. The optical response of metal nanoparticles (NPs) can be tuned by controlling their size, shape, and environment, providing a starting point for emerging research fields like surface plasmon-based photonics or plasmonics.

On the other hand, new synthesis methods developed to fabricate nanoparticles (NPs) with a specific size and shape enable us to tailor SP properties for clearly defined applications.

For instance, SPs are now being investigated for their potential applications in optics, magneto-optic, photonics, as a nanoengineered substrate on which the surface-enhanced Raman scattering response can be precisely controlled and optimized, and for chemical and biosensing applications, such as optical addressable diagnostic methods, devices, and therapies based on the plasmonic response of NPs.

A clear and general interpretation of SPR in metal NPs is not available today. For instance, the one to one correspondence between the SP resonances and the morphology of realistic NPs is still awaiting. In recent years, the influence of the geometry on the optical properties of metal nanoparticles has been an active research field. Several studies have shown that the main optical features depend on geometry and size, and the optical response of spherical, spheroidal, cubic, and other geometrical shapes, like rods and triangular prisms, are now well identified. However, it is known that metal nanoparticles of few nanometres show different structural motifs, such as icosahedra, octahedra, cuboctahedra, decahedra, etc., depending on their size, composition, and energetic conditions.

Furthermore, NPs can present both crystalline and noncrystalline arrangements and sometimes showing chiral structures.

Additionally to the technological implications, the optical signature of the NPs is also used as a characterization tool. The latter is interesting because optical techniques are nondestructive, and with proper implementation, they can be used to perform

measurements in situ and in real time, providing statistical properties of the whole sample. These attributes of optical spectroscopies are important because the properties of nanoparticles depend on the environment, and when growth and characterization are made in different conditions, the environment might be an additional uncontrollable variable for the optical interpretation. Thus, optical spectroscopies can be used also as complementary tools of structural characterization techniques like atomic force microscopy (AFM), scanning tunneling microscopy (STM), transmission electron microscopy (TEM), etc., which provide the image of a small piece of the sample, giving information about local properties and characterizing few NPs at a time.

Surface Plasmon Resonance

Surface plasmon resonance (SPR) is the collective oscillation of electrons in a solid or liquid stimulated by incident light. The resonance condition is established when the frequency of light photons matches the natural frequency of surface electrons oscillating against the restoring force of positive nuclei. SPR in nanometre-sized structures is called localized surface plasmon resonance.

SPR is the basis of many standard tools for measuring adsorption of material onto planar metal (typically gold and silver) surfaces or onto the surface of metal nanoparticles. It is the fundamental principle behind many colour-based biosensor applications and different lab-on-a-chip sensors.

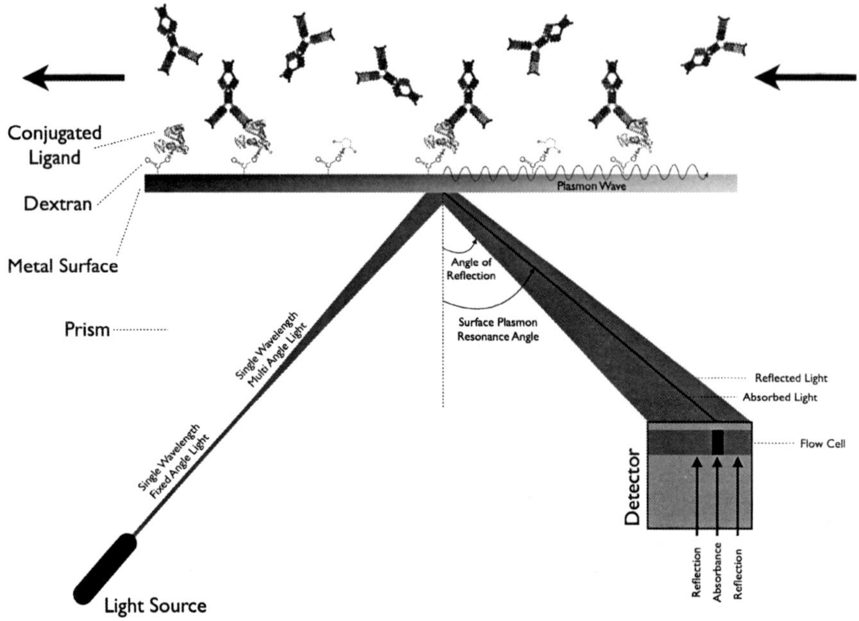

Figure: *Surface plasmon resonance (SPR).*

Explanation

Surface plasmon polaritons are surface electromagnetic waves that propagate in a direction parallel to the metal/dielectric (or metal/vacuum) interface. Since the wave is on the boundary of the metal and the external medium (air or water for example), these oscillations are very sensitive to any change of this boundary, such as the adsorption of molecules to the metal surface.

To describe the existence and properties of surface plasmon polaritons, one can choose from various models (quantum theory, Drude model, etc.). The simplest way to approach the problem is to treat each material as a homogeneous continuum, described by a frequency-dependent relative permittivity between the external medium and the surface. This quantity, hereafter referred to as the materials' "dielectric constant," is complex permittivity. In order for the terms which describe the electronic surface plasmons to exist, the real part of the dielectric constant of the metal must be negative and its magnitude must be greater than that of the dielectric. This condition is met in the IR-visible wavelength region for air/metal and water/metal interfaces (where the real dielectric constant of a metal is negative and that of air or water is positive).

LSPRs (Localized SPRs) are collective electron charge oscillations in metallic nanoparticles that are excited by light. They exhibit enhanced near-field amplitude at the resonance wavelength.

This field is highly localized at the nanoparticle and decays rapidly away from the nanoparticle/dieletric interface into the dielectric background, though far-field scattering by the particle is also enhanced by the resonance.

Light intensity enhancement is a very important aspect of LSPRs and localization means the LSPR has very high spatial resolution (subwavelength), limited only by the size of nanoparticles. Because of the enhanced field amplitude, effects that depend on the amplitude such as magneto-optical effect are also enhanced by LSPRs.

Realisation

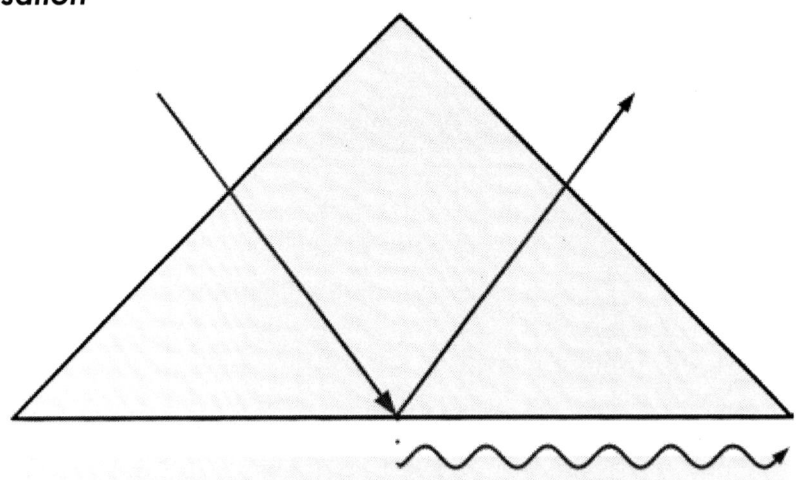

Figure: *Otto configuration*

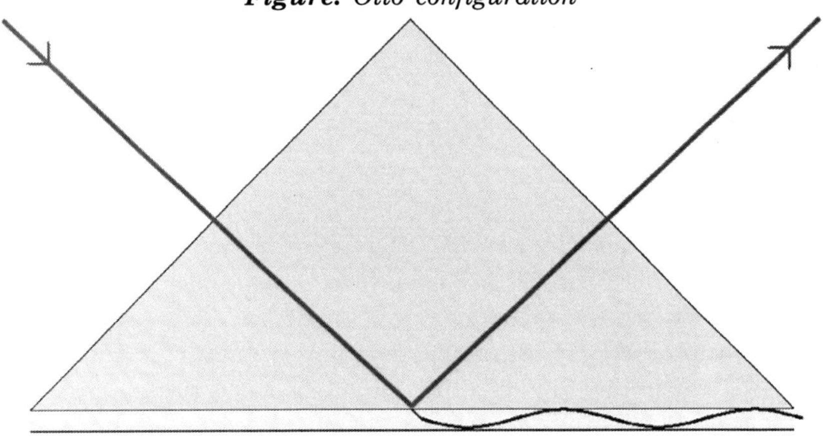

Figure: *Kretschmann configuration*

In order to excite surface plasmons in a resonant manner, one can use an electron or light beam (visible and infrared are typical). The incoming beam has to match its momentum to that of the plasmon. In the case of p-polarized light (polarization occurs parallel to the plane of incidence), this is possible by passing the light through a block of glass to increase the wavenumber (and the momentum), and achieve the resonance at a given wavelength and angle. S-polarized light (polarization occurs perpendicular to the plane of incidence) cannot excite electronic surface plasmons. Electronic and magnetic surface plasmons obey the following dispersion relation:

$$K(\omega) = \frac{\omega}{c}\sqrt{\frac{\varepsilon_1\varepsilon_2\mu_1\mu_2}{\varepsilon_1\mu_1 + \varepsilon_2\mu_2}}$$

where ϵ is the dielectric constant, and μ is the magnetic permeability of the material (1: the glass block, 2: the metal film).

Typical metals that support surface plasmons are silver and gold, but metals such as copper, titanium or chromium have also been used.

When using light to excite SP waves, there are two configurations which are well known. In the Otto setup, the light illuminates the wall of a glass block, typically a prism, and is totally internally reflected. A thin metal film (for example gold) is positioned close enough to the prism wall so that an evanescent wave can interact with the plasma waves on the surface and hence excite the plasmons.

In the Kretschmann configuration, the metal film is evaporated onto the glass block. The light again illuminates the glass block, and an evanescent wave penetrates through the metal film. The plasmons are excited at the outer side of the film. This configuration is used in most practical applications.

SPR Emission

When the surface plasmon wave interacts with a local particle or irregularity, such as a rough surface, part of the energy can be re-emitted as light. This emitted light can be detected *behind* the metal film from various directions.

Applications

Surface plasmons have been used to enhance the surface sensitivity of several spectroscopic measurements including fluorescence, Raman scattering, and second harmonic generation. However, in their simplest form, SPR reflectivity measurements can be used to detect molecular adsorption, such as polymers, DNA or proteins, etc. Technically, it is common, that the angle of the reflection minimum (absorption maximum) is measured. This angle changes in the order of 0.1° during thin (about nm thickness) film adsorption. In other cases the changes in the absorption wavelength is followed. The mechanism of detection is based on that the adsorbing molecules cause changes in the local index of refraction, changing the resonance conditions of the surface plasmon waves.

For nanoparticles, localized surface plasmon oscillations can give rise to the intense colours of suspensions or sols containing the nanoparticles. Nanoparticles or nanowires of noble metals exhibit strong absorption bands in the ultraviolet-visible light regime that are not present in the bulk metal. This extraordinary absorption increase has been exploited to increase light absorption in photovoltaic cells by depositing metal nanoparticles on the cell surface. The energy (colour) of this absorption differs when the light is polarized along or perpendicular to the nanowire. Shifts in this resonance due to changes in the local index of refraction upon adsorption to the nanoparticles can also be used to detect biopolymers such as DNA or proteins. Related complementary techniques include plasmon waveguide resonance, QCM, extraordinary optical transmission, and dual polarization interferometry

SPR Immunoassay

The first SPR immunoassay was proposed in 1983 by Liedberg, Nylander, and Lundström, then of the Linköping Institute of Technology (Sweden). They adsorbed human IgG onto a 600-angstrom silver film, and used the assay to detect anti-human IgG in water solution. Unlike many other immunoassays, such as ELISA, an SPR immunoassay is *label free* in that a label molecule is not required for detection of the analyte.

Data Interpretation

The most common data interpretation is based on the Fresnel formulas, which treat the formed thin films as infinite, continuous dielectric layers. This interpretation may result multiple possible refractive index and thickness values. However, usually only one solution is within the reasonable data range.

Metal particle plasmons are usually modelled using the Mie scattering theory. In many cases no detailed models are applied, but the sensors are calibrated for the specific application, and used with interpolation within the calibration curve.

Examples

Layer-by-Layer Self-Assembly

One of the first common applications of surface plasmon resonance spectroscopy was the measurement of the thickness (and refractive index) of adsorbed self-assembled nanofilms on gold substrates. The resonance curves shift to higher angles as the thickness of the adsorbed film increases. This example is a 'static SPR' measurement.

When higher speed observation is desired, one can select an angle right below the resonance point (the angle of minimum reflectance), and measure the reflectivity changes at that point. This is the so-called 'dynamic SPR' measurement. The interpretation of the data assumes that the structure of the film does not change significantly during the measurement.

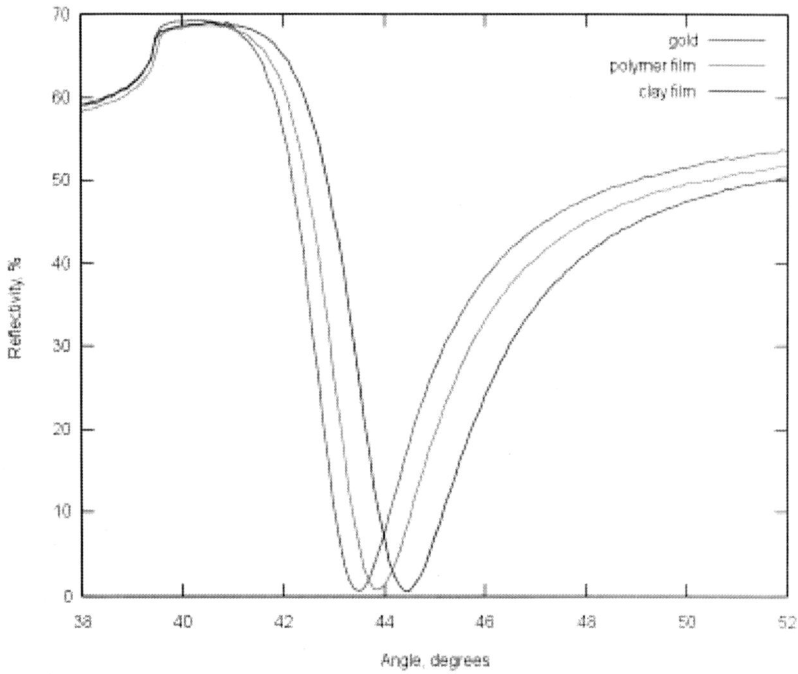

Figure: *SPR curves measured during the adsorption of a polyelectrolyte and then a clay mineral self-assembled film onto a thin (ca. 38 nanometres) gold sensor.*

Binding Constant Determination

When the affinity of two ligands has to be determined, the binding constant must be determined. It is the equilibrium value for the product quotient. This value can also be found using the dynamical SPR parameters and, as in any chemical reaction, it is the association rate divided by the dissociation rate.

For this, a so-called bait ligand is immobilized on the dextran surface of the SPR crystal. Through a microflow system, a solution with the prey analyte is injected over the bait layer. As the prey analyte binds the bait ligand, an increase in SPR signal (expressed in response units, RU) is observed. After desired association time, a solution without the prey analyte (usually the buffer) is injected on the microfluidics that dissociates the bound complex between bait ligand and prey analyte.

Now as the prey analyte dissociates from the bait ligand, a decrease in SPR signal (expressed in resonance units, RU) is observed. From these association ('on rate', k_a) and dissociation rates ('off rate', k_d), the equilibrium dissociation constant ('binding constant', K_D) can be calculated.

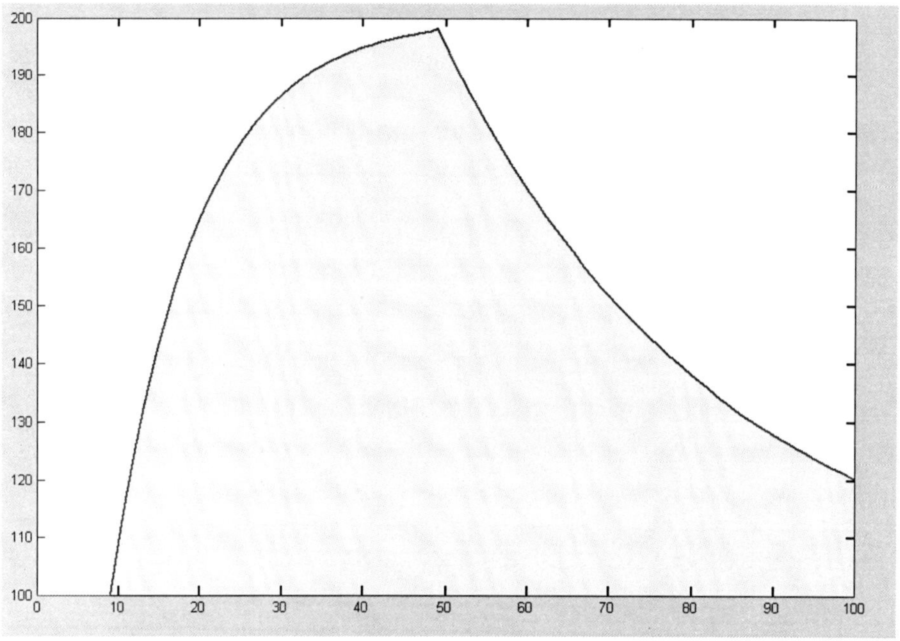

***Figure:** Association and dissociation signal*

The actual SPR signal can be explained by the electromagnetic 'coupling' of the incident light with the surface plasmon of the gold layer.

This plasmon can be influenced by the layer just a few nanometre across the gold–solution interface i.e. the bait protein and possibly the prey protein. Binding makes the reflection angle change;

$$K_D = \frac{k_d}{k_a}$$

Magnetic Plasmon Resonance

Recently, there has been an interest in magnetic surface plasmons. These require materials with large negative magnetic permeability, a property that has only recently been made available with the construction of metamaterials.

Application of Surface Plasmon Resonance Based on a Metal Nanoparticle

Several researchers have focused on biosensor for monitoring of biological interaction. Moreover, the detection of bimolecular is an extremely significant problem. Hence, the development of optical biosensors based on optical properties of noble metal nanoparticles using Surface Plasmon Resonance was considered.

Surface plasmon resonance (SPR) is a powerful technique to retrieve information on optical properties of biomaterial and nanomaterials. Biosensor based on SPR is a versatile technique for biological analysis applications. Essentially, SPR depends on the optical properties of metal layer and enviromental changes so it is related to charge density oscillation at the interface between them. Hence, biomolecular possess an extreme sensitivity to plasmon resonance and they remove the requirement for extrinsic biomolecular labelling. One advantage of SPR is, the light beam never passes through the dielectric medium of interest and hence the effect of absorption of the light in the analyte can be ignored. Hence, the main potential of surface plasmon resonance is haracterization of medium after the metal layer. Sometimes polymer layer was coated on gold later for detection and improvement the sensor performance. So, the properties of nanostructures gave rised to interest in biosensing.

Nano particles and Nano tube have improved the sensitivity, selectivity, and multiplexing capacity in biosensors. Different types of nanostructure-based biosensors including carbon nanotube, 2SnO nanoparticles, nanowires, and nanogap biosensors have emerged. Biomolecular interaction are determined and predicted via angular modulation sensor; hence, the out put of SPR biosensor sensor is angle shift (èspr), associated with the point of minimum reflected light intensity and is very sensitive to the changes in the dielectric constant of the medium. So, the sensitivity of modern SPR sensing systems based on the Kretschmann configuration is such that they are capable of detecting refractive indexes a high as approximately $5 \times 10^{"7}$ refractive index units, which corresponds to a 1 pg/mm^2 surface coverage of biomolecules.

Unfortunately, when sensing biomolecules of low molecular weight or in low concentration, the variation in SPR angle is very small and this causes difficulties in detecting the acquired signal [6]. Hence, several researchers have focused on enhancement of sensitivity of biomolecular sensor. Hence, the properties of metal nanoparticles (gold and silver) have contemplate as a result of the physical characteristics of surface-enhanced Raman scattering

(SERS). Zhu at al., utilised colloidal gold nanoparticle films to enhance the Raman scattering of the underlying molecules. On the other hand the colloidal gold nanoparticles were used as a immobilized element on the thin gold film. Moreover, to improve the roughness of the film, the inclusion of colloidal Au nanoparticles was added to the gold film. SPR biosensors using colloidal Au-modified gold films has the large angle shift and increase in minimum reflectivity suggesting that SPR spectrum depend on size and volume fraction of the Au nanoparticles. To improve the detection of SPR biosensors,

He et al., and; Lyon et al., reported using biomolecule-tagged colloidal Au nanoparticles. In this case, the SPW damped strongly because the intraction of the biomolecules -colloidal Au were occured with the immobilized bio-monolayer sensing. However, using this method to generate a biomolecular interaction signal involves complex procedures for the preparation of nano-sized colloidal Au particles and chemical immobilizations to tag the biomolecules. This process, eliminates the major advantage of

SPR technology, namely that biomolecular labelling is not required. On the methods for improving the sensitivity of SPR biosensors was reported by Gestwicki et. al.. They enhanced the sensitivity and response of biosensor based on absorption spectroscopy techniques. They shown the resonance angle can be improved by caused by the conformational changes of proteins and enzymes when they bind with analytes of low molecular weights and the response of sensor can be increased by adopting Au nanoshells or nanoclusters.

The Au nanocluster-embedded dielectric film is fabricated through a simple co-sputtering method utilising a multi-target sputtering system. In the proposed method, the resolution of the SPR biosensor is enhanced by precisely regulating the size and volume fraction of the

Au nanoclusters in order to control the surface plasmon effect.

Carbon nanotubes (CNTs) have been used as both electrode and transducer components in biosensors because of their interesting electrochemical and electrical properties including their flexibility for varying the electrical property (from metallic to semiconductor) and their, small diameters, high mechanical strength, and easy fabrication/integration. Wang [16] and Balasubramanian and Burghard, presented the design methodologies, performance characteristics, and potential applications of CNT biosensors, as well as the factors limiting their practical use. It is obvious that many challenges still do remain, especially regarding the sensitivity of the nanotube response to environmental effects, and also the chosen type of raw CNT material, and its functionalization.

The other nanostructured sensors is based on nano ceramic such as ZnO and 2SnO. For example, the 2SnO sensor has demonstrated a potential for gas sensing applications. Liu et al. developed a highly sensitive and nanostructured 2SnO thin-film sensor for the detection of ethanol vapor. Its lower LOD (below 1 ppm) demonstrates an improvement over that of other works reported in the literature. Recently, Choi et al. implemented a functionalized ZnO nanowire field effect transistor (FET) biosensor for the low level detection of biomolecular interactions. This biosensor was fabricated without any doping which showed an enhanced sensitivity compared to other nanowire biosensors.

Fundamentally, the base of the biosensor is the coupling of a ligand-receptor binding reaction to a signal transducer. Moreover, angular, phase and polarization modulations are the various methods applicable for distinguishing the interaction of bimolecular with the receptor. Essentially, SPR is a quantum electromagnetic phenomenon that appears at the interface of the dielectric and the metal. Under certain conditions, the energy of the light beam is absorbed by collective excitation of the free electrons called surface Plasmon (SP), which lies between the interface of the dielectric and the metal. On the other hand, when the momentum of the photon matches that of the Plasmon, the resonance appears as an interface of two media with dielectric constants of opposite signs, and the SP wave propagates along the interface. In accordance with the SP wave properties, the SPR is classified as propagating the SPR, the long-range SPR and the localized SPR.

In the case of propagating SPR (PSPR), the amplitude of the SP wave is decaying exponentially normal to the interface, which the Plasmon propagates, for a distance of about one to ten microns. The interactions between the surface and molecules in the analyte lead to the SPR signal shift which can be observed in shift of the angle of resonance and the wavelength. A second type of Plasmon's wave is Long-range SPR (LRSPR) that occurs in thin metal films with low attenuation; hence, the travelling distance of the Plasmon wave is about a millimetre in the visible range.

The localized SPR (LSPR) is exited on a nano-scale. On the other hand, the LSPR appears in the metal structure as a lateral size that is less than half the wavelength of the excitation photon. Hence, the light beam will interacts with the nanoparticles, so as to leads to a locally Plasmon oscillation occurring around the particle with a specific frequency similar to LSPR. However, the LSPR is more sensitive to changes in the local dielectric analyte. Typically, researchers will measure the changes in the local dielectric environment or analyte via making shifts in the LSPR wavelength measurements. Moreover a variant of angle-resolved sensing that is possible for LSPR is also possible, and LSPR can provide real-time kinetic data for the binding processes. In addition, the PSPR signal is more sensitive to changes in the bulk refractive index than are the LSPR signals, so when the

LSPR and PSPR signals are examined the interaction with the molecular adsorption layer and, the response of the two methods becomes comparable in the short-range. Consequently, both sensitivity and flexibility are the capability of the LSPR. Moreover, the decay length is shorter than PSPR; hence, the LSPR are offered for the sensor.

Optical Properties of the Metal Nanoparticles

The nanoparticles of interest are typically large enough to accurately apply classical electromagnetic theory to describe their interaction with light. However, they are also small enough to allow observation strong variations in the optical properties depending on the particle size, shape, and local environment. Because of the complexity of the systems being studied, efficient computational methods are the capable of treating large size materials are essential. In the past few years, several numerical methods have been developed to determine the optical properties of small particles, including the discrete dipole approximation (DDA) [23], T-matrix, and spectral representation methods (SR) [25].

We employed the DDA, which is a computational procedure suitable for studying the scattering and absorption of electromagnetic radiation by particles with sizes in the order or less of the wavelength than the incident light. DDA has been applied to a broad range of problems, including interstellar and interplanetary dust grains, ice crystals in the atmosphere, human blood cells, surface features of semiconductors, metal nanoparticles and their aggregates, and more. The DDA was first introduced by Purcell and Pennypacker and has been subjected to several improvements.

The main idea behind DDA is to approximate a scatter, in our case the nanoparticle, by large enough array of polarizable point dipoles. Once the location and polarizability of each dipole are specified, the calculation of the scattering and absorption efficiencies by the dipole array can be determined, depending on the accuracy of the mathematical algorithms and the capabilities of the computational hardware. Although the calculation of the radiated fields in DDA is, in theory and principle, also possible, that calculation is actually beyond the computational capabilities of any of the current systems.

4

Magnetic Clusters and Nanoparticles

In physics, the term clusters denotes small, multiatom particles. As a rule of thumb, any particle of somewhere between 3 and 3×10^7 atoms is considered a cluster. Two-atom particles are sometimes considered clusters as well..The term can also refer to the organisation of protons and neutrons within an atomic nucleus, e.g. the Alpha particle (also known as "α-cluster"), consisting of two protons and two neutrons (as in a helium nucleus).

Although first reports of cluster species date back already to the 1940s, Cluster science emerged as a separate direction of research in the 1980s, One purpose of the research was to study the gradual development of collective phenomena which characterize a bulk solid. These are for example the colour of a body, its electrical conductivity, its ability to absorb or reflect light, and magnetic phenomena such as ferro-, ferri-, or antiferromagnetism. These are typical collective phenomena which only develop in an aggregate of a large number of atoms.

It was found that collective phenomena break down for very small cluster sizes. It turned out, for example, that small clusters of a ferromagnetic material are super-paramagnetic rather than ferromagnetic. Paramagnetism is not a collective phenomenon, which means that the ferromagnetism of the macrostate was not conserved by going into the nanostate. The question then was asked for example, "How many atoms do we need in order to obtain the collective metallic or magnetic properties of a solid?" Soon after the first cluster sources had been developed in 1980, an ever larger community of cluster scientists was involved in such studies. This development led to the discovery of fullerenes in 1986 and carbon nanotubes a few years later.

In science, a lot is known about properties of the gas phase; however, comparatively little is known about the condensed phases (the liquid phase and solid phase.) The study of clusters attempts to bridge this gap of knowledge by clustering atoms together and studying their characteristics. If enough atoms were clustered together, eventually one would obtain a liquid or solid.

The study of atomic and molecular clusters also benefits the developing field of nanotechnology. If new materials are to be made out of nanoscale particles, such as nanocatalysts and quantum computers, the properties of the nanoscale particles (the clusters) must first be understood.

Magnetic clusters, i.e., assemblies of a finite number (between two or three and several hundred) of interacting spin centres which are magnetically decoupled from their environment, can be found in many materials ranging from inorganic compounds, magnetic molecules, artificial metal structures formed on surfaces to metalloproteins. The magnetic excitation spectra in them are determined by the nature of the spin centres, the nature of the magnetic interactions, and the particular arrangement of the mutual interaction paths between the spin centres. Small clusters of up to four magnetic ions are ideal model systems to examine the fundamental magnetic interactions which are usually dominated by Heisenberg exchange, but often complemented by anisotropic and/or higher-order interactions. In large magnetic clusters which may potentially deal with a dozen or more spin centres, the possibility of novel many-body quantum states and quantum phenomena are in focus. In this review the necessary theoretical concepts and experimental techniques to study the magnetic cluster excitations and the resulting characteristic magnetic properties are introduced, followed by examples of small clusters demonstrating the enormous amount of detailed physical information which can be retrieved. The current understanding of the excitations and their physical interpretation in the molecular nanomagnets which represent large magnetic clusters is then presented, with an own section devoted to the subclass of the single-molecule magnets which are distinguished by displaying quantum tunneling of the magnetization.

The ability to shape matter on the nanoscale, i.e. in the size regime *between* the molecular and the microscopic levels is a major technological challenge of this century. A possible way to accomplish this goal is to assemble such structures from chemically modified inorganic particles.

These particles can, depending on the functional groups on their surfaces, undergo specific linkage reactions to from complex particle aggregates, whose structures are determined by the selective chemical interactions between the particles and by their relative sizes. This self-assembly process can lead to discrete superstructures of infinite variety and with function.

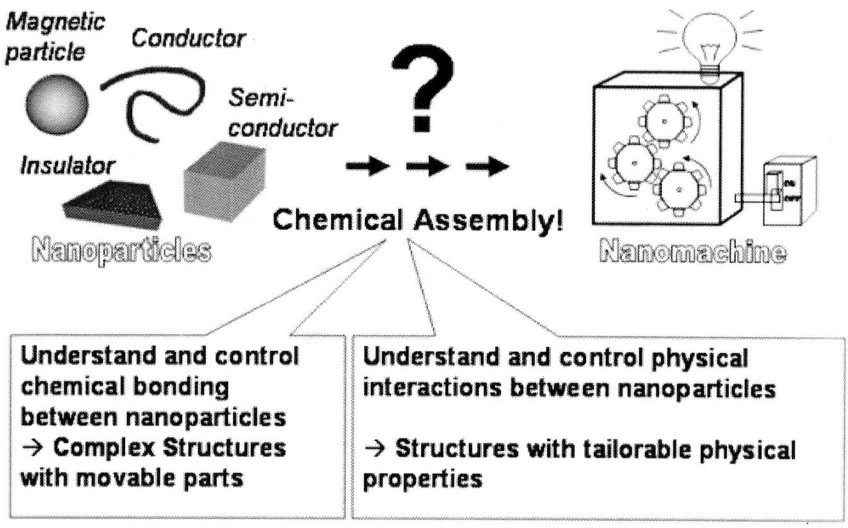

Research in the group is devoted to developing selective methods for the linkage of inorganic nanoparticles into larger aggregates and to investigating the bonding and physical properties of the resulting structures. Examples for nanoparticle-based structures with zero-, one- and two-dimensional geometries are shown below. These nanoparticle clusters are supported by covalent or electrostatic interactions between the nanoparticles. They form by reaction of inorganic nanoparticles functionalized with organic linkers their surfaces. By adjusting the nanoparticle surface chemistries and the particle stoichiometry, it is possible to control the structures and bonding in these clusters.

Magnetic nanoparticles (MNPs) have many potential biomedical applications. Improvements in their magnetic properties and solubility are necessary for these applications to realise their full potential.

Single-Molecule Magnet

Single-molecule magnets or SMMs are a class of metalorganic compounds, that show superparamagnetic behaviour below a certain blocking temperature at the molecular scale. In this temperature range, SMMs exhibit magnetic hysteresis of purely molecular origin. Contrary to conventional bulk magnets and molecule-based magnets, collective long-range magnetic ordering of magnetic moments is not necessary.

Intramolecular Coupling

The magnetic coupling between the spins of the metal ions is mediated via superexchange interactions and can be described by the following isotropic Heisenberg Hamiltonian:

$$\hat{\mathcal{H}}_{HB} = \sum_{i<j} J_{i,j} \mathrm{S}_i \cdot \mathrm{S}_j$$

where $J_{i,j}$ is the coupling constant between spin i (operator S_i) and spin j (operator S_j). For positive J the coupling is called ferromagnetic (parallel alignment of spins) and for negative J the coupling is called antiferromagnetic (antiparallel alignment of spins).

- a high spin ground state,
- a high zero-field-splitting (due to high magnetic anisotropy), and
- negligible magnetic interaction between molecules.

The combination of these properties can lead to an energy barrier so that, at low temperatures, the system can be trapped in one of the high-spin energy wells.

"These molecules contain a finite number of interacting spin centres (e.g. paramagnetic ions) and thus provide ideal opportunities to study basic concepts of magnetism. Some of them possess magnetic ground states and give rise to hysteresis effects and metastable magnetic phases. They may show quantum tunneling of the magnetization which raises the question of coherent dynamics in such systems. Other types of molecules exhibit pronounced frustration effects, whereas so-called spin crossover substances can switch their magnetic ground state and related properties such as colour under irradiation of laser light, pressure or heat. Scientists from various fields – chemistry, physics; theory and experiment – have joined the research on molecular magnetism in order to explore the unprecedented properties of these new compounds."

"Single-molecule magnets (SMMs) have many important advantages over conventional nanoscale magnetic particles composed of metals, metal alloys or metal oxides. These advantages include uniform size, solubility in organic solvents, and readily alterable peripheral ligands, among others."

"A single molecule magnet is an example of a macroscopic quantum system. [...] If we could detect spin flips in a single atom or molecule, we could use the spin to store information. This would enable us to increase the storage capacity of computer hard disks. [...] A good starting point for trying to detect spin flips is to find a molecule with a spin of several Bohr magnetons. [An electron has an intrinsic magnetic dipole moment of approximately one Bohr magneton.] There is a very well studied molecular magnet, Mn_{12}-acetate, which has a spin S = 10. This molecule is a disc-shaped organic molecule in which twelve Mn ions are embedded. Eight of these form a ring, each having a charge of +3 and a spin S = 2.

The other four form a tetrahedron, each having a charge of +4 and a spin S = 3/2. The exchange interactions within the molecule are such that the spins of the ring align themselves in opposition to the spins of the tetrahedron, giving the molecule a total net spin S = 10."

Blocking Temperature

Measurements take place at very low temperatures. The so-called blocking temperature is defined as the temperature below which the relaxation of the magnetisation becomes slow compared to the time scale of a particular investigation technique. A molecule magnetised at 2 K will keep 40% of its magnetisation after 2 months and by lowering the temperature to 1.5 K this will take 40 years.

Future Applications

Figure: *One possible use of SMMs is superior magnetic thin films to coat hard disks.*

As of 2008 there are many discovered types and potential uses. "Single molecule magnets (SMM) are a class of molecules exhibiting magnetic properties similar to those observed in conventional bulk magnets, but of molecular origin. SMMs have been proposed as potential candidates for several technological applications that require highly controlled thin films and patterns."

"The ability of a single molecule to behave like a tiny magnet (single molecular magnets, SMMs) has seen a rapid growth in research over the last few years. SMMs represent the smallest possible magnetic devices and are a controllable, bottom-up approach to nanoscale magnetism. Potential applications of SMMs include quantum computing, high-density information storage and magnetic refrigeration."

"A single molecule magnet is an example of a macroscopic quantum system. [...] If we could detect spin flips in a single atom or molecule, we could use the spin to store information. This would enable us to increase the storage capacity of computer hard disks. [...] A good starting point for trying to detect spin flips is to find a molecule with a spin of several Bohr magnetons. [An electron has an intrinsic magnetic dipole moment of approximately one Bohr magneton.] There is a very well studied molecular magnet, Mn_{12}-acetate, which has a spin S = 10. This molecule is a disc-shaped organic molecule in which twelve Mn ions are embedded.

Eight of these form a ring, each having a charge of +3 and a spin S = 2. The other four form a tetrahedron, each having a charge of +4 and a spin S = 3/2. The exchange interactions within the molecule are such that the spins of the ring align themselves in opposition to the spins of the tetrahedron, giving the molecule a total net spin S = 10."

Types

The archetype of single-molecule magnets is called "Mn_{12}". It is a polymetallic manganese (Mn) complex having the formula $[Mn_{12}O_{12}(OAc)_{16}(H_2O)_4]$, where OAc stands for acetate. It has the remarkable property of showing an extremely slow relaxation of their magnetization below a blocking temperature. $[Mn_{12}O_{12}(OAc)_{16}(H_2O)_4]\cdot 4H_2O\cdot 2AcOH$ which is called "Mn_{12}-acetate" is a common form of this used in research.

"Mn_4" is another researched type single-molecule magnet. Three of these are:

- $[Mn_4(hmp)_6(NO_3)_2(MeCN)_2](ClO_4)_2\cdot 2MeCN$
- $[Mn_4(hmp)_6(NO_3)_4]\cdot(MeCN)$
- $[Mn_4(hmp)_4(acac)_2(MeO)_2](ClO_4)_2\cdot 2MeOH$

In each of these Mn_4 complexes "there is a planar diamond core of $Mn^{III}{}_2Mn^{II}{}_2$ ions. An analysis of the variable-temperature and variable-field magnetization data indicate that all three molecules have intramolecular ferromagnetic coupling and a S = 9 ground state. The presence of a frequency-dependent alternating current susceptibility signal indicates a significant energy barrier between the spin-up and spin-down states for each of these three $Mn^{III}{}_2Mn^{II}{}_2$ complexes."

Single-molecule magnets are also based on iron clusters because they potentially have large spin states. In addition the biomolecule ferritin is also considered a nanomagnet. In the cluster Fe_8Br the cation Fe_8 stands for $[Fe_8O_2(OH)_{12}(tacn)_6]^{8+}$ with tacn representing 1,4,7-triazacyclononane.

The ferrous cube complex $Fe_4C_{40}H_{52}N_4O_{12}$ (commonly called $[Fe_4(sae)_4(MeOH)_4]$) was the first example of a single-molecule magnet involving an Fe(II) cluster, and the core of this complex is a slightly distorted cube with Fe and O atoms on alternating corners. Remarkably, this single molecule magnet exhibits non-collinear magnetism in which the atomic spin moments of the four Fe atoms point in opposite directions along two nearly perpendicular axes. Theoretical computations showed approximately two magnetic electrons are localized on each Fe atom with the other atoms being nearly nonmagnetic, and the spin-orbit coupling potential energy surface has three local energy minima with a magnetic anisotropy barrier just below 3 meV.

History

Although the term "single-molecule magnet" was first employed in 1996, the first single-molecule magnet was reported in 1991. The complex $Mn_{12}O_{12}(MeCO_2)_{16}(H_2O)_4$ complex ($Mn_{12}Ac_{16}$), first described in 1980, exhibits slow relaxation of the magnetization at low temperatures. This manganese oxide compound features a central $Mn(IV)_4O_4$ cube surrounded by a ring of 8 Mn(III) units connected through bridging oxo ligands.

It was known in 2006 that the "deliberate structural distortion of a Mn_6 compound via the use of a bulky salicylaldoxime derivative switches the intra-triangular magnetic exchange from antiferromagnetic to ferromagnetic resulting in an S = 12 ground state.

A record magnetization was reported in 2007 for $[Mn(III)_6O_2(sao)_6(O2CPh)_2(EtOH)_4]$, with S = 12, D = -0.43 cm^{-1} and hence U = 62 cm^{-1} or 86 K at a blocking temperature of 4.3 K. This was accomplished by replacing acetate ligands (OAc) by the bulkier salicylaldoxime thus distorting the manganese ligand sphere. It is prepared by mixing the perchlorate of manganese, the sodium salt of benzoic acid, a salicylaldoxime derivate and tetramethylammonium hydroxide in water and collecting the filtrate.

In 2011 it was reported by the University of Nottingham that a dinuclear complex of depleted uranium could be mistaken for a single molecule magnet by chemist Stephen Liddle.

Detailed Behaviour

Molecular magnets exhibit an increasing product (magnetic susceptibility times temperature) with decreasing temperature, and can be characterized by a shift both in position and intensity of the a.c. magnetic susceptibility. Single-molecule magnets represent a molecular approach to nanomagnets (nanoscale magnetic particles). In addition, single-molecule magnets have provided physicists with useful test-beds for the study of quantum mechanics. Macroscopic quantum tunneling of the magnetization was first observed in $Mn_{12}O_{12}$, characterized by evenly-spaced steps in the hysteresis curve. The periodic quenching of this tunneling rate in the compound Fe_8 has been observed and explained with geometric phases.

Due to the typically large, bi-stable spin anisotropy, single-molecule magnets promise the realisation of perhaps the smallest practical unit for magnetic memory, and thus are possible building blocks for a quantum computer. Consequently, many groups have devoted great efforts into synthesis of additional single molecule magnets; however, the $Mn_{12}O_{12}$ complex and analogous complexes remain the canonical single molecule magnet with a 50 cm^{-1} spin anisotropy.

The spin anisotropy manifests itself as an energy barrier that spins must overcome when they switch from parallel alignment to antiparallel alignment. This barrier (U) is defined as:

$$U = S^2 \mid D \mid$$

where S is the dimensionless total spin state and D the zero-field splitting parameter (in cm^{-1}); D can be negative but only its absolute value is considered in the equation. The barrier U is generally reported in cm^{-1} units or in units of Kelvin (see: electronvolt). The higher the barrier the longer a material remains magnetized and a high barrier is obtained when the molecule contains many unpaired electrons and when its zero field splitting value is large. For example, the "$Mn(OAc)_{12}$" cluster the spin state is 10 (involving 20 unpaired electrons) and D = -0.5 cm^{-1} resulting in a barrier of 50 cm^{-1} (equivalent to 60 K).

The effect is also observed by hysteresis experienced when magnetization is measured in a magnetic field sweep: on lowering the magnetic field again after reaching the maximum magnetization the magnetization remains at high levels and it requires a reversed field to bring magnetization back to zero. Recently, it has been reported that the energy barrier, U, is slightly dependent on Mn_{12} crystal size/ morphology, as well as the magnetization relaxation times, which varies as function of particle size and size distributions.

Ferromagnetism

Ferromagnetism is the basic mechanism by which certain materials (such as iron) form permanent magnets, or are attracted to magnets. In physics, several different types of magnetism are distinguished. Ferromagnetism (including ferrimagnetism) is the strongest type; it is the only type that creates forces strong enough to be felt, and is responsible for the common phenomena of magnetism encountered in everyday life.

Other substances respond weakly to magnetic fields with two other types of magnetism, paramagnetism and diamagnetism, but the forces are so weak that they can only be detected by sensitive instruments in a laboratory. An everyday example of ferromagnetism is a refrigerator magnet used to hold notes on a refrigerator door. The attraction between a magnet and ferromagnetic material is "the quality of magnetism first apparent to the ancient world, and to us today".

Permanent magnets (materials that can be magnetized by an external magnetic field and remain magnetized after the external field is removed) are either ferromagnetic or ferrimagnetic, as are other materials that are noticeably attracted to them. Only a few substances are ferromagnetic. The common ones are iron, nickel, cobalt and most of their alloys, some compounds of rare earth metals, and a few naturally-occurring minerals such as lodestone.

Ferromagnetism is very important in industry and modern technology, and is the basis for many electrical and electromechanical devices such as electromagnets, electric motors, generators, transformers, and magnetic storage such as tape recorders, and hard disks.

Historically, the term *ferromagnet* was used for any material that could exhibit spontaneous magnetization: a net magnetic moment in the absence of an external magnetic field. This general definition is still in common use. More recently, however, different classes of spontaneous magnetization have been identified when there is more than one magnetic ion per primitive cell of the material, leading to a stricter definition of "ferromagnetism" that is often used to distinguish it from ferrimagnetism. In particular,

- a material is "ferromagnetic" in this narrower sense only if *all* of its magnetic ions add a positive contribution to the net magnetization.
- If some of the magnetic ions *subtract* from the net magnetization (if they are partially *anti*-aligned), then the material is "ferrimagnetic".

- If the moments of the aligned and anti-aligned ions balance completely so as to have zero net magnetization, despite the magnetic ordering, then it is an antiferromagnet. These alignment effects only occur at temperatures below a certain critical temperature, called the Curie temperature (for ferromagnets and ferrimagnets) or the Néel temperature (for antiferromagnets).

Among the first investigations of ferromagnetism are the pioneering works of Aleksandr Stoletov on measurement of the magnetic permeability of ferromagnetics, known as the Stoletov curve.

Ferromagnetic Materials

Table: *Curie temperatures for some crystalline ferromagnetic (* = ferrimagnetic) materials*

Material	*Curietemp. (K)*
Co	1388
Fe	1043
Fe_2O_3*	948
$FeOFe_2O_3$*	858
$NiOFe_2O_3$*	858
$CuOFe_2O_3$*	728
$MgOFe_2O_3$*	713
MnBi	630
Ni	627
MnSb	587
$MnOFe_2O_3$*	573
$Y_3Fe_5O_{12}$*	560
CrO_2	386
MnAs	318
Gd	292
Dy	88
EuO	69

The table on the right lists a selection of ferromagnetic and ferrimagnetic compounds, along with the temperature above which they cease to exhibit spontaneous magnetization.

Ferromagnetism is a property not just of the chemical make-up of a material, but of its crystalline structure and microscopic organisation. There are ferromagnetic metal alloys whose constituents are not themselves ferromagnetic, called Heusler alloys, named after Fritz

Heusler. Conversely there are non-magnetic alloys, such as types of stainless steel, composed almost exclusively of ferromagnetic metals.

One can also make amorphous (non-crystalline) ferromagnetic metallic alloys by very rapid quenching (cooling) of a liquid alloy. These have the advantage that their properties are nearly isotropic (not aligned along a crystal axis); this results in low coercivity, low hysteresis loss, high permeability, and high electrical resistivity. One such typical material is a transition metal-metalloid alloy, made from about 80% transition metal (usually Fe, Co, or Ni) and a metalloid component (B, C, Si, P, or Al) that lowers the melting point.

A relatively new class of exceptionally strong ferromagnetic materials are the rare-earth magnets. They contain lanthanide elements that are known for their ability to carry large magnetic moments in well-localized f-orbitals.

Curie Temperature

In physics and materials science, the Curie temperature (T_c), or Curie point, is the temperature where a material's permanent magnetism changes to induced magnetism. The force of magnetism is determined by magnetic moments.

The Curie temperature is the critical point where a material's intrinsic magnetic moments change direction. Magnetic moments are permanent dipole moments within the atom which originate from electrons' angular momentum and spin. Materials have different structures of intrinsic magnetic moments that depend on temperature. At a material's Curie Temperature those intrinsic magnetic moments change direction. Permanent magnetism is caused by the alignment of magnetic moments and induced magnetism is created when disordered magnetic moments are forced to align in an applied magnetic field. For example, the ordered magnetic moments (ferromagnetic) change and become disordered (paramagnetic) at the Curie Temperature.

Higher temperatures make magnets weaker as spontaneous magnetism only occurs below the Curie Temperature. Magnetic susceptibility only occurs above the Curie Temperature and can be calculated from the Curie-Weiss Law which is derived from Curie's Law.

In analogy to ferromagnetic and paramagnetic materials, the Curie temperature can also be used to describe the temperature where a material's spontaneous electric polarisation changes to induced electric polarisation or the reverse upon reduction of the temperature below the Curie temperature.

The Curie temperature is named after Pierre Curie who showed that magnetism was lost at a critical temperature.

Magnetic Moments

Magnetic moments are permanent dipole moments within the atom which are made up from electrons angular momentum and spin.

Electrons inside atoms contribute magnetic moments from their own angular momentum and from their orbital momentum around the nucleus. Magnetic moments from the nucleus are insignificant in contrast to magnetic moments from electrons. Thermal contribution will result in higher energy electrons causing disruption to their order and alignment between dipoles to be destroyed.

Ferromagnetic, paramagnetic, ferrimagnetic and antiferromagnetic materials have different structures of intrinsic magnetic moments. It is at a material's specific Curie Temperature where they change properties. The transition from antiferromagnetic to paramagnetic (or vice versa) occurs at the Néel Temperature which is analogous to Curie Temperature.

Below T_c		**Above T_c**
Ferromagnetic	↔	Paramagnetic
Ferrimagnetic	↔	Paramagnetic
Antiferromagnetic	↔	Paramagnetic

Orientations of Magnetic Moments in Materials

Figure: *Ferromagnetism The magnetic moments in a ferromagnetic material. The moments are ordered and of the same magnitude in the absence of an applied magnetic field.*

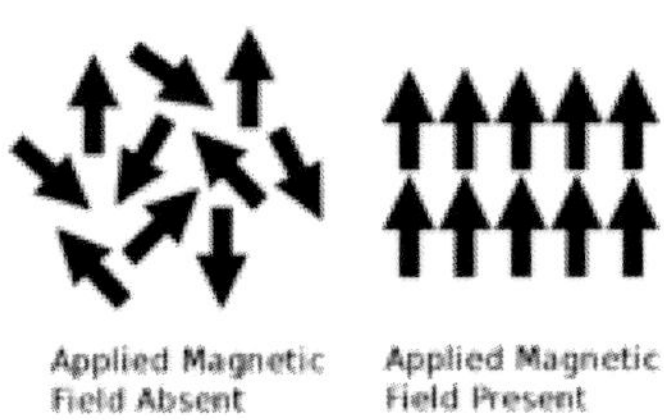

Figure: *Paramagnetism The magnetic moments in a paramagnetic material. The moments are disordered in the absence of an applied magnetic field and ordered in the presence of an applied magnetic field.*

Figure: *Ferrimagnetism The magnetic moments in a ferrimagnetic material. The moments are aligned oppositely and have different magnitudes due to being made up of two different ions. This is in the absence of an applied magnetic field.*

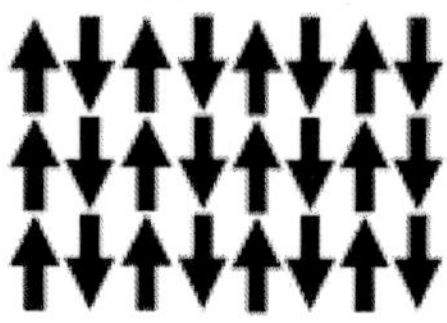

Figure: *Antiferromagnetism The magnetic moments in an antiferromagnetic material. The moments are aligned oppositely and have the same magnitudes. This is in the absence of an applied magnetic field.*

Materials with Magnetic Moments that Change Properties at the Curie Temperature

Ferromagnetic, paramagnetic, ferrimagnetic and antiferromagnetic structures are made up of intrinsic magnetic moments. If all electrons within the structure are paired, these moments cancel out due to having opposite spins and angular momentum. Thus even with an applied magnetic field will have different properties and no Curie Temperature.

Paramagnetic

A material is paramagnetic only above its Curie Temperature. Paramagnetic materials are non-magnetic when a magnetic field is absent and magnetic when a magnetic field is applied. When the magnetic field is absent the material has disordered magnetic moments; that is, the atoms are unsymmetrical and not aligned. When the magnetic field is present the magnetic moments are temporarily realigned parallel to the applied field; the atoms are symmetrical and aligned. The magnetic moment in the same direction is what causes an induced magnetic field.

For paramagnetism this response to an applied magnetic field is positive and known as magnetic susceptibility. The magnetic susceptibility only applies above the Curie Temperature for disordered states.

Sources of Paramagnetism (Materials which have Curie Temperatures);

- All atoms which have unpaired electrons;
- Atoms where inner shells are incomplete in electrons;
- Free radicals;
- Metals

Above the Curie Temperature the atoms are excited, the spin orientation becomes randomised, but can be realigned in an applied field and the material paramagnetic. Below the Curie Temperature the intrinsic structure has under gone a phase transition, the atoms are ordered and the material is ferromagnetic. The paramagnetic materials induced magnetic fields are very weak in comparison to ferromagnetic materials magnetic fields.

Ferromagnetic

Materials are only ferromagnetic below their corresponding Curie temperatures. Ferromagnetic materials are magnetic in the absence of an applied magnetic field. When a magnetic field is absent the material has spontaneous magnetization which is a result of the ordered magnetic moments; that is, for ferromagnetism, the atoms are symmetrical and aligned in the same direction creating a permanent magnetic field.

The magnetic interactions are held together by exchange interactions; otherwise thermal disorder would overcome the weak interactions of magnetic moments. The exchange interaction has a zero probability of parallel electrons occupying the same point in time, implying a preferred parallel alignment in the material. The Boltzmann factor contributes heavily as it prefers interacting particles to be aligned in the same direction. This causes ferromagnets to have strong magnetic fields and high Curie temperatures of around 1000K.

Below the Curie temperature, the atoms are aligned and parallel, causing spontaneous magnetism; the material is ferromagnetic. Above the Curie temperature the material is paramagnetic, as the atoms lose their ordered magnetic moments when the material undergoes a phase transition.

Ferrimagnetic

Not to be confused with ferromagnetic.

Materials are only ferrimagnetic below their materials corresponding Curie Temperature. Ferrimagnetic materials are magnetic in the absence of an applied magnetic field and are made up of two different ions.

When a magnetic field is absent the material has a spontaneous magnetism which is the result of ordered magnetic moments; that is, for ferrimagnetism one ion's magnetic moments are aligned facing in one direction with certain magnitude and the other ion's magnetic moments are aligned facing in the opposite direction with a different magnitude. As the magnetic moments are of different magnitudes in opposite directions there is still a spontaneous magnetism and a magnetic field is present.

Similar to ferromagnetic materials the magnetic interactions are held together by exchange interactions. The orientations of moments however are anti-parallel which results in a net momentum by subtracting their momentum from one another.

Below the Curie Temperature the atoms of each ion are aligned anti-parallel with different momentums causing a spontaneous magnetism; the material is ferrimagnetic. Above the Curie Temperature the material is paramagnetic as the atoms lose their ordered magnetic moments as the material undergoes a phase transition.

Antiferromagnetic and the Néel Temperature

Materials are only antiferromagetic below their corresponding Néel Temperature. This is similar to the Curie Temperature as above the Néel Temperature the material undergoes a phase transition and becomes paramagnetic.

The material has equal magnetic moments aligned in opposite directions resulting in a zero magnetic moment and a net magnetism of zero at all temperatures below the Néel Temperature. Antiferromagnetic materials are weakly magnetic in the absence or presence of an applied magnetic field.

Similar to ferromagnetic materials the magnetic interactions are held together by exchange interactions preventing thermal disorder from overcoming the weak interactions of magnetic moments. When disorder occurs it is at the Néel Temperature.

Curie-Weiss Law

The Curie-Weiss law is an adapted version of Curie's law.

The Curie-Weiss law is a simple model derived from a mean-field approximation, this means it works well for the materials temperature, T, much greater than their corresponding Curie Temperature, T_c, i.e. $T >> T_c$; however fails to describe the magnetic susceptibility, χ, in the immediate vicinity of the Curie point because of local fluctuations between atoms.

Both Curie's law and the Curie-Weiss law do not hold for $T < T_c$.

Curie's law for a paramagnetic material;

$$\chi = \frac{M}{H} = \frac{M\mu_0}{B} = \frac{C}{T}$$

Definition	
χ	the magnetic susceptibility; the influence of an applied magnetic field on a material
M	the magnetic moments per unit volume
H	the macroscopic magnetic field
B	the magnetic field
C	the material-specific Curie constant

$$C = \frac{\mu_0 \mu_B^2}{3k_B} N g^2 J(J+1)$$

μ_0	the permeability of free space. Note - in CGS units is taken to equal one.
g	the Landé g-factor
J(J+1)	the eigenvalue for eigenstate J^2 for the stationary states within the incomplete atoms shells (electrons unpaired)
μ_B	the Bohr Magneton
k_B	Boltzmann's constant
total magnetism	is N number of magnetic moments per unit volume

The Curie-Weiss law is then derived from Curie's law to be

$$\chi = \frac{C}{T - T_c}$$

where

$$T_C = \frac{C\lambda}{\mu_0}$$

λ is the Weiss molecular field constant.

Physics of Curie Temperature

Approaching Curie Temperature from Above

As the Curie-Weiss Law is an approximation a more accurate model is needed when the temperature,T, approaches the materials Curie Temperature,T_C.

Magnetic susceptibility occurs above the Curie Temperature.

An accurate model of critical behaviour for magnetic susceptibility with critical exponent γ;

$$\chi \sim \frac{1}{(T-T_c)^{\gamma}}$$

The critical exponent differs between materials and for the mean-field model is taken as $\gamma=1$.

As temperature is inversely proportional to magnetic susceptibility when T approaches T_C the denominator tends to zero and the magnetic susceptibility approaches infinity allowing magnetism to occur. This is a spontaneous magnetism which is a property of ferromagnetic and ferrimagnetic materials.

Approaching Curie Temperature from Below

Magnetism depends on temperature and spontaneous magnetism occurs below the Curie Temperature. An accurate model of critical behaviour for spontaneous magnetism with critical exponent β;

$$M \sim (T-T_C)^{\beta}$$

The critical exponent differs between materials and for the mean-field model as taken as $\beta=0.5$ where $T \ll T_C$.

The spontaneous magnetism approaches zero as the temperature increases towards the materials Curie Temperature.

Approaching absolute zero (0 Kelvin)

The spontaneous magnetism, occurring in ferromagnetic, ferrimagnetic and antiferromagnetic materials, approaches zero as the temperature increases towards the material's Curie Temperature. Spontaneous magnetism is at its maximum as the temperature approaches 0K. That is, the magnetic moments are completely aligned and at their strongest magnitude of magnetism due to no thermal disturbance.

In paramagnetic materials temperature is sufficient to overcome the ordered alignments. As the temperature approaches 0K the entropy decreases to zero, that is, the disorder decreases and becomes ordered. This occurs without the presence of an applied magnetic field and obeys the third law of thermodynamics.

Both Curie's Law and the Curie-Weiss law fail as the temperature approaches 0K. This is because they depend on the magnetic susceptibility which only applies when the state is disordered.

Gadolinium Sulphate continues to satisfy Curie's law at 1K. Between 0-1K the law fails to hold and a sudden change in the intrinsic structure occurs at the Curie Temperature.

Ising Model of Phase Transitions

The Ising model is mathematically based and can analyse the critical points of phase transitions in ferromagnetic order due to spins of electrons having magnitudes of either +/- ½. The spins interact with their neighbouring dipole electrons in the structure and here the Ising model can predict their behaviour with each other.

This model is important for solving and understanding the concepts of phase transitions and hence solving the Curie Temperature. As a result many different dependencies that effect the Curie Temperature can be analysed.

For example the surface and bulk properties depend on the alignment and magnitude of spins and the Ising model can determine the effects of magnetism in this system.

Weiss Domains and Surface and Bulk Curie Temperatures

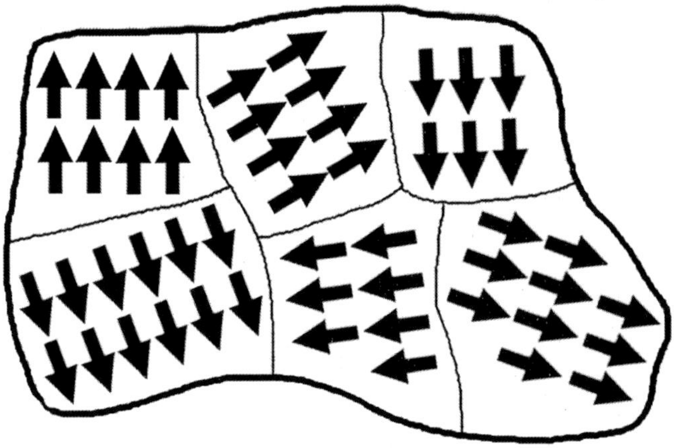

Figure: *The Weiss domains in a ferromagnetic material; the magnetic moments are aligned in domains.*

Materials structures consist of intrinsic magnetic moments which are separated into domains called Weiss domains. This can result in ferromagnetic materials having no spontaneous magnetism as domains could potentially balance each other out. The position of particles can therefore have different orientations around the surface than the main part (bulk) of the material. This property directly affects the Curie Temperature as there can be a bulk Curie Temperature T_B and a different surface Curie Temperature T_S for a material.

This allows for the surface Curie Temperature to be ferromagnetic above the bulk Curie Temperature when the main state is disordered, i.e. Ordered and disordered states occur simultaneously.

The surface and bulk properties can be predicted by the Ising model and electron capture spectroscopy can be used to detect the electron spins and hence the magnetic moments on the surface of the material. An average total magnetism is taken from the bulk and surface temperatures to calculate the Curie Temperature from the material, noting the bulk contributes more.

The angular momentum of an electron is either +'/2 or - '/2 due to it having a spin of ½, which gives a specific size of magnetic moment to the electron; the Bohr Magneton. Electrons orbiting around the nucleus in a current loop create a magnetic field which depends on the Bohr Magneton and magnetic quantum number. Therefore the magnetic moments are related between angular and orbital momentum and affect each other. Angular momentum contributes twice as much to magnetic moments than orbital.

For terbium which is a rare earth metal and has a high orbital angular momentum the magnetic moment is strong enough to affect the order above its bulk temperatures. It is said to have a high anisotropy on the surface, that is it is highly directed in one orientation. It remains ferromagnetic on its surface above its Curie Temperature while its bulk becomes ferrimagnetic and then at higher temperatures its surface remains ferrimagnetic above its bulk Néel Temperature before becoming completely disordered and paramagnetic with increasing temperature. The anisotropy in the bulk is different from its surface anisotropy just above these phase changes as the magnetic moments will be ordered differently or ordered in paramagnetic materials.

Changing a Material's Curie Temperature

Composite Materials

Composite materials, that is, materials composed from other materials with different properties, can change the Curie Temperature. For example a composite which has silver in can create spaces for oxygen molecules in bonding which decreases the Curie Temperature as the crystal lattice will not be as compact.

The alignment of magnetic moments in the composite material affects the Curie Temperature. If the materials moments are parallel with each other the Curie Temperature will increase and if perpendicular the Curie Temperature will decrease as either more or less thermal energy will be needed to destroy the alignments. Preparing composite materials through different temperatures can result in different final compositions which will have different Curie Temperatures. Doping a material can also affect its Curie Temperature.

The density of nanocomposite materials changes the Curie Temperature. Nanocomposites are compact structures on a nano-scale. The structure is built up of high and low bulk Curie Temperatures, however will only have one mean-field Curie Temperature. A higher density of lower bulk temperatures results in a lower mean-field Curie Temperature and a higher density of higher bulk temperature significantly increases the mean-field Curie Temperature. In more than one dimension the Curie Temperature begins to increase as the magnetic moments will need more thermal energy to overcome the ordered structure.

Particle Size

The size of particles in a material's crystal lattice changes the Curie Temperature. Due to the small size of particles (nanoparticles) the fluctuations of electron spins become more prominent, this results in the Curie Temperature drastically decreasing when the size of particles decrease as the fluctuations cause disorder. The size of a particle also affects the anisotropy causing alignment to become less stable and thus lead to disorder in magnetic moments.

The extreme of this is superparamagnetism which only occurs in small ferromagnetic particles and is where fluctuations are very influential causing magnetic moments to change direction randomly and thus create disorder.

The Curie Temperature of nanoparticles are also affected by the crystal lattice structure, body-centred cubic (bcc), face-centred cubic (fcc) and a hexagonal structure (hcp) all have different Curie Temperatures due to magnetic moments reacting to their neighbouring electron spins. fcc and hcp have tighter structures and as a results have higher Curie Temperatures than bcc as the magnetic moments have stronger effects when closer together. This is known as the coordination number which is the number of nearest neighbouring particles in a structure. This indicates a lower coordination number at the surface of a material than the bulk which leads to the surface becoming less significant when the temperature is approaching the Curie Temperature. In smaller systems the coordination number for the surface is more significant and the magnetic moments have a stronger affect on the system.

Although fluctuations in particles can be minuscule, they are heavily dependent on the structure of crystal lattices as they react with their nearest neighbouring particles. Fluctuations are also affected by the exchange interaction as parallel facing magnetic moments are favoured and therefore have less disturbance and disorder, therefore a tighter structure influences a stronger magnetism and therefore a higher Curie Temperature.

Pressure

Pressure changes a material's Curie Temperature. Increasing pressure on the crystal lattice decreases the volume of the system. Pressure directly affects the kinetic energy in particles as movement increases causing the vibrations to disrupt the order of magnetic moments. This is similar to temperature as it also increases the kinetic energy of particles and destroys the order of magnetic moments and magnetism.

Pressure also affects the density of states (DOS). Here the DOS decreases causing the number of electrons available to the system to decrease. This leads to the number of magnetic moments decreasing as they depend on electron spins. It would be expected because of this that the Curie Temperature would decrease however it increases. This is the result of the exchange interaction. The exchange interaction favours the aligned parallel magnetic moments due to electrons being unable to occupy the same space in time and as this is increased due to the volume decreasing the Curie Temperature increases with pressure. The Curie Temperature is made up of a combination of dependencies on kinetic energy and the DOS.

It is interesting to note that the concentration of particles also affects the Curie Temperature when pressure is being applied and can result in a decrease in Curie Temperature when the concentration is above a certain percent.

Orbital Ordering

Orbital ordering changes the Curie Temperature of a material. Orbital ordering can be controlled through applied Strains. This is a function that determines the wave of a single electron or paired electrons inside the material. Having control over the probability of where the electron will be allows the Curie Temperature to be altered. For example the delocalised electrons can be moved onto the same plane by applied strains within the crystal lattice.

The Curie Temperature is seen to increase greatly due to electrons being packed together in the same plane, they are forced to align due to the exchange interaction and thus increases the strength of the magnetic moments which prevents thermal disorder at lower temperatures.

Curie Temperature in Ferroelectric and Piezoelectric Materials

In analogy to ferromagnetic and paramagnetic materials, the Curie Temperature can also used to describe the temperature where a material's spontaneous electric polarisation changes to induced electric polarisation, or vice versa.

Electric polarisation is a result of aligned electric dipoles. Aligned electric dipoles are composites of positive and negative charges where all the dipoles are facing in one direction. The charges are separated from their stable placement in the particles and can occur spontaneously, from pressure or an applied electric field.

Ferroelectric, dielectric (paraelectric) and piezoelectric materials have electric polarisation. In ferroelectric materials there is a spontaneous electric polarisation in the absence of an applied electric field. In dielectric materials there is electric polarisation aligned only when an electric field is applied. Piezoelectric materials have electric polarisation due to applied mechanical stress distorting the structure from pressure.

T_0 is the temperature where ferroelectric materials lose their spontaneous polarisation as a first or second order phase change occurs, that is the internal structure changes or the internal symmetry changes. In certain cases T_0 is equal to the Curie Temperature however the Curie Temperature can be 10 kelvin lower than T_0.

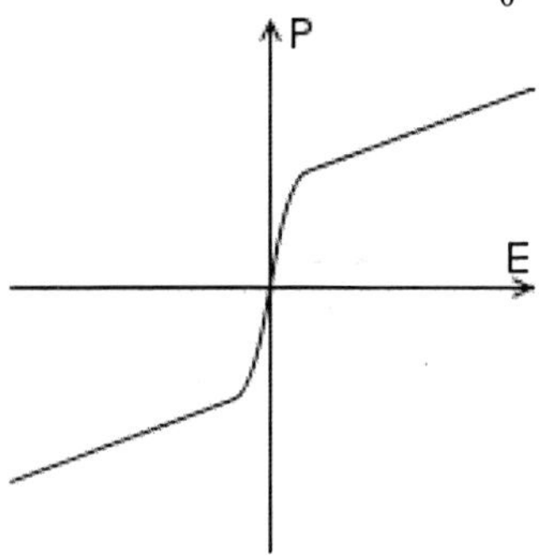

Figure: *(Below T_0) Ferroelectric polarisation P in an applied electric field E.*

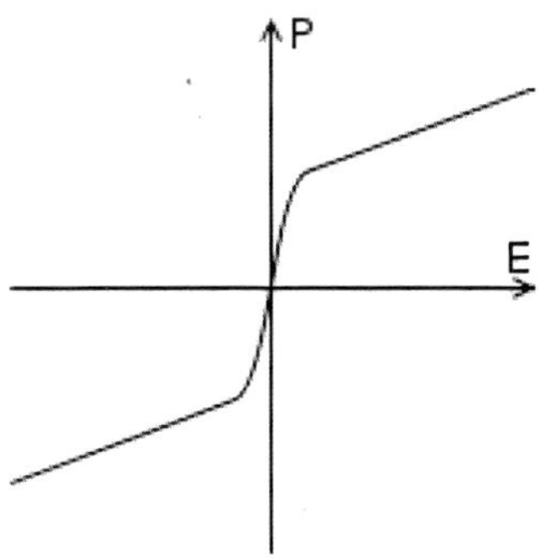

Figure: *(Above T_0) Dielectric polarisation P in an applied electric field E.*

Below T_0	***Above T_0***Ferroelectric	↔ Dielectric

All ferroelectric materials are piezoelectric.

Piezoelectric

An external force applies pressure on particles inside the material which affects the structure of the crystal lattice. Particles in a unit cell become unsymmetrical which allows a net polarisation from each particle. Symmetry would cancel the opposing charges out and there would be no net polarisation. Below the transition temperature T_0 displacement of electric charges causes polarisation. Above the transition temperature T_0 the structure is cubic and symmetric, causing the material to become dielectric. Electric charges are also agitated and disordered causing the material to have no electric polarisation in the absence of an applied electric field.

Ferroelectric and Dielectric

Materials are only ferroelectric below their corresponding transition temperature T_0. Ferroelectric materials are all piezoelectric and therefore have a spontaneous electric polarisation as the structures are unsymmetrical.

Materials are only dielectric above their corresponding transition temperature T_0. Dielectric materials have no electric polarisation in the absence of an applied electric field. The electric dipoles are unaligned and have no net polarisation. In analogy to magnetic susceptibility, electric susceptibility only occurs above T_0.

Ferroelectric materials when polarised are influenced under hysteresis; that is they are dependent on their past state as well as their current state. As an electric field is applied the dipoles are forced to align and polarisation is created, when the electric field is removed polarisation remains. The hysteresis loop depends on temperature and as a result as the temperature is increased and reaches T_0 the two curves become one curve as shown in the dielectric polarisation.

Relative Permittivity

A modified version of the Curie Weiss law applies to the dielectric constant, also known as the relative permittivity:

$$\epsilon = \epsilon_0 + \frac{C}{T - T_c}$$

Applications

A heat-induced ferromagnetic-paramagnetic transition is used in magneto-optical storage media, for erasing and writing of new data. Famous examples include the Sony Minidisc format, as well as the now-obsolete CD-MO format. Other uses include temperature control in soldering irons, and stabilizing the magnetic field of tachometre generators against temperature variation.

Actinide Ferromagnets

A number of actinide compounds are ferromagnets at room temperature or become ferromagnets below the Curie temperature (T_C). PuP is one actinide pnictide that is a paramagnet and has cubic symmetry at room temperature, but upon cooling undergoes a lattice distortion to tetragonal when cooled to below its T_c = 125 K. PuP has an easy axis of <100>, so that

$$\frac{c}{a}-1=-(31\pm1)\times10^{-4}$$

at 5 K. The lattice distortion is presumably a consequence of strain induced by the magnetoelastic interactions as the magnetic moments aligned parallel within magnetic domains.

In $NpFe_2$ the easy axis is <111>. Above T_C ~500 K $NpFe_2$ is also paramagnetic and cubic. Cooling below the Curie temperature produces a rhombohedral distortion wherein the rhombohedral angle changes from 60° (cubic phase) to 60.53°. An alternate description of this distortion is to consider the length c along the unique trigonal axis (after the distortion has begun) and a as the distance in the plane perpendicular to c. In the cubic phase this reduces to $\frac{c}{a}$ = 1.00. Below the Curie temperature

$$\frac{c}{a}-1=-(120\pm5)\times10^{-4}$$

which is the largest strain in any actinide compound. $NpNi_2$ undergoes a similar lattice distortion below T_C = 32 K, with a strain of (43 ± 5) × 10^{-4}. $NpCo_2$ is a ferrimagnet below 15 K.

Lithium Gas

In 2009, a team of MIT physicists demonstrated that a lithium gas cooled to less than one kelvin can exhibit ferromagnetism. The team cooled fermionic lithium-6 to less than 150 billionths of one kelvin above absolute zero using infrared laser cooling. This demonstration is the first time that ferromagnetism has been demonstrated in a gas.

Explanation

The Bohr–van Leeuwen theorem shows that magnetism cannot occur in purely classical solids. Without quantum mechanics, there would be no diamagnetism, paramagnetism or ferromagnetism. The property of ferromagnetism is due to the direct influence of two effects from quantum mechanics: spin and the Pauli exclusion principle.

Origin of Magnetism

One of the fundamental properties of an electron (besides that it carries charge) is that it has a magnetic dipole moment, i.e., it behaves itself as a tiny magnet. This dipole moment comes from the more fundamental property of the electron that it has quantum mechanical spin. The quantum mechanical nature of this spin causes the electron to only be able to be in two states, with the magnetic field either pointing "up" or "down" (for any choice of up and down). The spin of the electrons in atoms is the main source of ferromagnetism, although there is also a contribution from the orbital angular momentum of the electron about the nucleus. When these tiny magnetic dipoles are aligned in the same direction, their individual magnetic fields add together to create a measurable macroscopic field.

However, in materials with a filled electron shell, the total dipole moment of the electrons is zero because the spins are in up/down pairs. Only atoms with partially filled shells (i.e., unpaired spins) can have a net magnetic moment, so ferromagnetism only occurs in materials with partially filled shells. Because of Hund's rules, the first few electrons in a shell tend to have the same spin, thereby increasing the total dipole moment.

These unpaired dipoles (often called simply "spins" even though they also generally include angular momentum) tend to align in parallel to an external magnetic field, an effect called paramagnetism. Ferromagnetism involves an additional phenomenon, however: The dipoles tend to align spontaneously, giving rise to a spontaneous magnetization, even when there is no applied field.

Exchange Interaction

According to classical electromagnetism, two nearby magnetic dipoles will tend to align in *opposite* directions, so their magnetic fields will oppose one another and cancel out. However, this effect is very weak, because the magnetic fields generated by individual spins are small and the resulting alignment is easily destroyed by thermal fluctuations. In a few materials, a much stronger interaction between spins arises because the change in the direction of the spin leads to a change in electrostatic repulsion between neighbouring electrons, due to a particular quantum mechanical effect called the exchange interaction. At short distances, the exchange interaction is much stronger than the dipole-dipole magnetic interaction. As a result, in a few materials, the ferromagnetic ones, nearby spins tend to align in the same direction. In certain doped semiconductor oxides RKKY interactions have been shown to bring about periodic longer-range magnetic interactions, a phenomenon of significance in the study of spintronic materials.

The exchange interaction is related to the Pauli exclusion principle, which says that two electrons with the same spin cannot also have the same "position". Therefore, under certain conditions, when the orbitals of the unpaired outer valence electrons from adjacent atoms overlap, the distributions of their electric charge in space are further apart when the electrons have parallel spins than when they have opposite spins. This reduces the electrostatic energy of the electrons when their spins are parallel compared to their energy when the spins are anti-parallel, so the parallel-spin state is more stable. In simple terms, the electrons, which repel one another, can move "further apart" by aligning their spins, so the spins of these electrons tend to line up. This difference in energy is called the exchange energy.

The materials in which the exchange interaction is much stronger than the competing dipole-dipole interaction are frequently called *magnetic materials*. For instance, in iron (Fe) the exchange force is about 1000 times stronger than the dipole interaction. Therefore below the Curie temperature virtually all of the dipoles in a ferromagnetic material will be aligned. The exchange interaction is also responsible for the other types of spontaneous ordering of atomic magnetic moments occurring in magnetic solids, antiferromagnetism and ferrimagnetism. There are different exchange interaction mechanisms which create the magnetism in different ferromagnetic, ferrimagnetic, and antiferromagnetic substances. These mechanisms include direct exchange, RKKY exchange, double exchange, and superexchange.

Magnetic Anisotropy

Although the exchange interaction keeps spins aligned, it does not align them in a particular direction. Without magnetic anisotropy, the spins in a magnet randomly change direction in response to thermal fluctuations and the magnet is superparamagnetic. There are several kinds of magnetic anisotropy, the most common of which is magnetocrystalline anisotropy. This is a dependence of the energy on the direction of magnetization relative to the crystallographic lattice. Another common source of anisotropy, inverse magnetostriction, is induced by internal strains. Single-domain magnets also can have a *shape anisotropy* due to the magnetostatic effects of the particle shape. As the temperature of a magnet increases, the anisotropy tends to decrease, and there is often a blocking temperature at which a transition to superparamagnetism occurs.

Magnetic Domains

The above would seem to suggest that every piece of ferromagnetic material should have a strong magnetic field, since all the spins are aligned, yet iron and other ferromagnets are often found in an "unmagnetized" state. The reason for this is that a bulk piece of ferromagnetic material is divided into tiny *magnetic domains* (also known as *Weiss domains*).

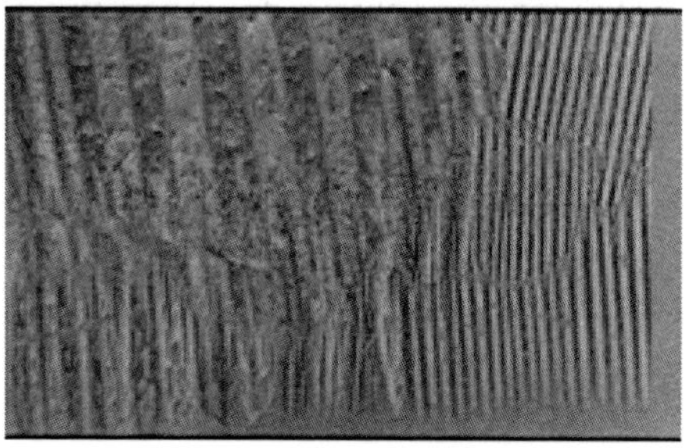

Figure: *Electromagnetic dynamic magnetic domain motion of grain oriented electrical silicon steel*

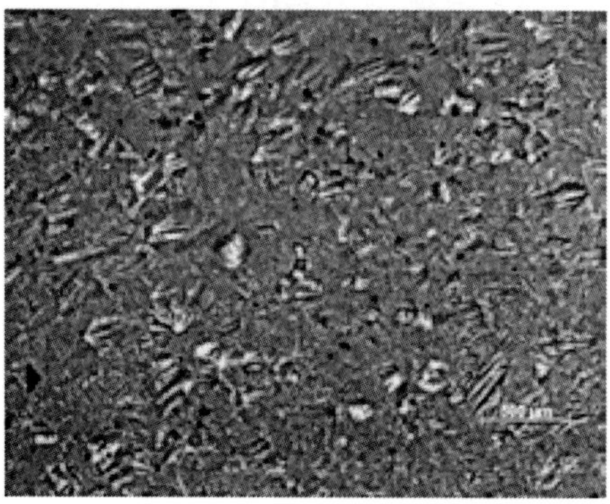

Figure: *Domain orientations in non-grain-oriented steel imaged with magneto-optical sensor and polarizer microscope.*

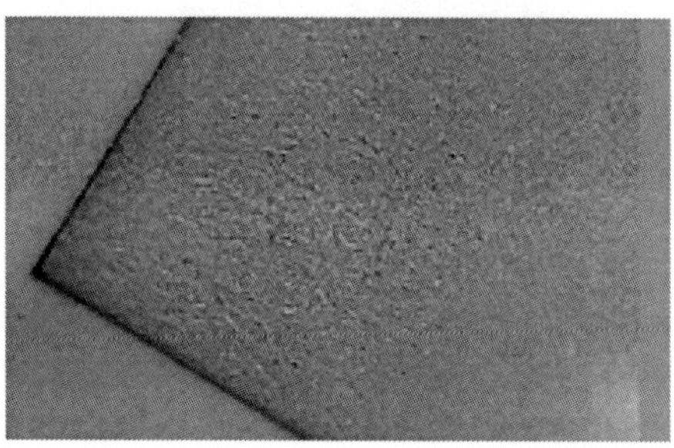

Figure: *Magnetic domains of non-oriented silicon or electrical steel (recorded with CMOS-MagView).*

Figure: *Kerr micrograph of metal surface showing magnetic domains. The domains are the red and green stripes within each microcrystalline grain. The magnetic field in the red domains is in the opposite direction from the green domains.*

Within each domain, the spins are aligned, but (if the bulk material is in its lowest energy configuration, i.e. *unmagnetized*), the spins of separate domains point in different directions and their magnetic fields cancel out, so the object has no net large scale magnetic field.

Ferromagnetic materials spontaneously divide into magnetic domains because the *exchange interaction* is a short-range force, so over long distances of many atoms the tendency of the magnetic dipoles to reduce their energy by orienting in opposite directions wins out. If all the dipoles in a piece of ferromagnetic material are aligned parallel, it creates a large magnetic field extending into the space around it. This contains a lot of magnetostatic energy. The material can reduce this energy by splitting into many domains pointing in different directions, so the magnetic field is confined to small local fields in the material, reducing the volume of the field. The domains are separated by thin domain walls a number of molecules thick, in which the direction of magnetization of the dipoles rotates smoothly from one domain's direction to the other.

Magnetized Materials

Thus, a piece of iron in its lowest energy state ("unmagnetized") generally has little or no net magnetic field. However, if it is placed in a strong enough external magnetic field, the domain walls will move, reorienting the domains so more of the dipoles are aligned with the external field. The domains will remain aligned when the external field is removed, creating a magnetic field of their own extending into the space around the material, thus creating a "permanent" magnet. The domains do not go back to their original minimum energy configuration when the field is removed because the domain walls tend to become 'pinned' or 'snagged' on defects in the crystal lattice, preserving their parallel orientation. This is shown by the Barkhausen effect: as the magnetizing field is changed, the magnetization changes in thousands of tiny discontinuous jumps as the domain walls suddenly "snap" past defects. This magnetization as a function of the external field is described by a hysteresis curve. Although this state of aligned domains found in a piece of magnetized ferromagnetic material is not a minimal-energy configuration, it is metastable, and can persist for long periods, as shown by samples of magnetite from the sea floor which have maintained their magnetization for millions of years.

Alloys used for the strongest permanent magnets are "hard" alloys made with many defects in their crystal structure where the domain walls "catch" and stabilize. The net magnetization can be destroyed by heating and then cooling (annealing) the material without an external field, however. The thermal motion allows the domain boundaries to move, releasing them from any defects, to return to their low-energy unaligned state.

Curie Temperature

As the temperature increases, thermal motion, or entropy, competes with the ferromagnetic tendency for dipoles to align. When the temperature rises beyond a certain point, called the Curie temperature, there is a second-order phase transition and the system can no longer maintain a spontaneous magnetization, although it still responds paramagnetically to an external field. Below that temperature, there is a spontaneous symmetry breaking and magnetic moments become aligned with their neighbours. The Curie temperature itself is a critical point, where the magnetic susceptibility is theoretically infinite and, although there is no net magnetization, domain-like spin correlations fluctuate at all length scales.

The study of ferromagnetic phase transitions, especially via the simplified Ising spin model, had an important impact on the development of statistical physics. There, it was first clearly shown that mean field theory approaches failed to predict the correct behaviour at the critical point (which was found to fall under a *universality class* that includes many other systems, such as liquid-gas transitions), and had to be replaced by renormalization group theory.

Single-Molecule Experiment

A single-molecule experiment is an experiment that investigates the properties of individual molecules. Single-molecule studies may be contrasted with measurements on an ensemble or bulk collection of molecules, where the individual behaviour of molecules cannot be distinguished, and only average characteristics can be measured. Since many measurement techniques in biology, chemistry and physics are not sensitive enough to observe single molecules, single-molecule fluorescence techniques (that have emerged since the 90s for probing various processes on the level of individual molecules) caused a lot of excitement, since these supplied many new details on the measured processes that were not accessible in the past. Indeed, since the 90s, many techniques for probing individual molecules were developed.

The first single-molecule experiments were patch clamp experiments performed in the 70s, but these were limited to studying ion channels. Today, systems investigated using single-molecule techniques include the movement of myosin on actin filaments in muscle tissue and the spectroscopic details of individual local environments in solids. Biological polymers' conformations have been measured using atomic force microscopy (AFM). Using force spectroscopy, single molecules (or pairs of interacting molecules), usually polymers, can be mechanically stretched and their elastic response recorded in real time.

In the gas phase at ultralow pressures, single-molecule experiments have been around for decades, but in the condensed phase only since 1989 with the work by W. E. Moerner and Lothar Kador. One year later Michel Orrit and Jacky Bernard were able to show also the detection of the absorption of single molecules by their fluorescence.

Many techniques have the ability to observe one molecule at a time, most notably mass spectrometry, where single ions are detected. In addition one of the earliest means of detecting single molecules, came about in the field of ion channels with the development of the patch clamp technique by Erwin Neher and Bert Sakmann (who later went on to win the Nobel prize for their seminal contributions) However, the idea of measuring conductance to look at single molecules placed a serious limitation on the kind of systems which could be observed.

Fluorescence is a convenient means of observing one molecule at a time, mostly due to the sensitivity of commercial optical detectors, capable of counting single photons.

However, spectroscopically, the observation of one molecule requires that the molecule is in an isolated environment and that it emits photons upon excitation, which owing to the technology to detect single photons by use of photomultiplier tubes (PMT) or avalanche photodiodes (APD), enables one to record photon emission events with great sensitivity and time resolution.

More recently, single molecule fluorescence is the subject of intense interest for biological imaging, through the labelling of biomolecules such as proteins and nucleotides to study enzymatic function which cannot easily be studied on the bulk scale, due to subtle time-dependent movcments in catalysis and structural reorganisation. The most studied protein has been the class of myosin/actin enzymes found in muscle tissues. Through single molecule techniques the step mechanism has been observed and characterized in many of these proteins.

Nanomanipulators such as the atomic force microscope are also suited to single molecule experiments of biological significance, since they work on the same length scale of most biological polymers. Besides, atomic force microscopy (AFM) is appropriate for the studies of synthetic polymer molecules. AFM provides a unique possibility of 3D visualization of polymer chains. For instance, AFM tapping mode is gentle enough for the recording of adsorbed polyelectrolyte molecules (for example, 0.4 nm thick chains of poly(2-vinylpyridine)) under liquid medium. The location of two-chain-superposition correspond in these experiments to twice the thickness of single chain (0.8 nm in the case of the mentioned example). At the application of proper scanning parameters, conformation of such molecules remain unchanged for hours that allows the performance of experiments under liquid media having various properties (Roiter and Minko, 2005). Optical tweezers have also been used with success.

About the Experiments

Concept

Single molecule fluorescence spectroscopy uses the fluorescence of a molecule for obtaining information on its environment, structure, and position. The technique affords the ability of obtaining information otherwise not available due to ensemble averaging (that is, a signal obtained when recording many molecules at the same time represents an average property of the molecules' dynamics). The results in many experiments of individual molecules are two-state trajectories.

Single Channel Recording

As in the case of single molecule fluorescence spectroscopy, the technique known as single channel recording can be used to obtain specific kinetic information—in this case about ion channel function—that is not available when ensemble recording, such as whole-cell recording, is performed. Specifically, ion channels alternate between conducting and non-conducting classes, which differ in conformation. Therefore, the functional state of ion channels can be directly measured with sufficiently sensitive electronics, provided that proper precautions are taken to minimize noise. In turn, each of these classes may be divided into one or more kinetic states with direct bearing on the underlying function of the ion channel. Performing these types of single molecule studies under systematically varying conditions (e.g. agonist concentration and structure, permeant ion and/or channel blocker, mutations in the ion channel amino acids), can provide information regarding the interconversion of various kinetic states of the ion channel. In a minimal model for an ion channel, there are two states: open and closed.

However, other states are often needed in order to accurately represent the data, including multiple closed states as well as inactive and/or desensitized states, which are non-conducting states that can occur even in the presence of stimulus.

Figure: *Two examples of data (~10 s each) from a single channel recording experiment using the patch clamp technique. The upper trace is at a lower concentration, while the lower trace is recorded at an agonist concentration near this ion channel's EC_{50}. Channel openings are represented by downward deflections, in this case deflections of approximately 5 pA.*

Biomolecule Labelling

Single fluorophores can be chemically attached to biomolecules, such as proteins or DNA, and the dynamics of individual molecules can be tracked by monitoring the fluorescent probe. Spatial movements within the Rayleigh limit can be tracked, along with changes in emission intensity and/or radiative lifetime, which often indicate changes in local environment. For instance, single-molecule labelling has yielded a vast quantity of information on how kinesin motor proteins move along microtubule strands in muscle cells.

Single-Molecule Fluorescence Resonance Energy Transfer

Here, the molecule is labelled in (at least) two places. A laser beam is focused on the molecule exciting the first probe. When this probe relaxes and emits a photon, it has a chance of exciting the other probe. The efficiency of the absorption of the photon emitted from the first probe in the second probe depends on the distance between these probes. Since the distance changes with time, this experiment probes the internal dynamics of the molecule.

Theoretical Notes

Single-molecule experiments versus ensemble experiments: When looking at data related to individual molecules, one usually can construct propagators, and jumping time probability density functions, of the first order, the second order and so on, whereas from bulk experiments, one usually obtains the decay of a correlation function. From the information contained in these unique functions (obtained from individual molecules), one can extract a relatively clear picture on the way the system behaves; e.g. its kinetic scheme, or its potential of activity, or its reduced dimensions form.

In particular, one can construct (many properties of) the reaction pathway of an enzyme when monitoring the activity of an individual enzyme. Additionally, significant aspects regarding the analysis of single molecule data—such as fitting methods and tests for homogeneous populations—have been described by several authors. On the other hand, there are several issues with the analysis of single molecule data including construction of a low noise environment and insulated pipet tips, filtering some of the remaining unwanted components (noise) found in recordings, and the length of time required for data analysis (pre-processing, unambiguous event detection, plotting data, fitting kinetic schemes, etc.).

Impact

Single-molecule techniques impacted optics, electronics, biology, and chemistry. In the biological sciences, the study of proteins and other complex biological machinery was limited to ensemble experiments that nearly made impossible the direct observation of their kinetics. For example, it was only after single molecule fluorescence microscopy was used to study kinesin-myosin pairs in muscle tissue that direct observation of the walking mechanisms were understood. These experiments, however, have for the most part been limited to in vitro studies, as useful techniques for live cell imaging have yet to be fully realised. The promise of single molecule in vivo imaging, however, brings with it an enormous potential to directly observe bio-molecules in native processes. These techniques are often targeted for studies involving low-copy proteins, many of which are still being discovered. These techniques have also been extended to study ares of chemistry, including the mapping of heterogeneous surfaces.

Magnetic Anisotropy

Magnetic anisotropy is the directional dependence of a material's magnetic properties. In the absence of an applied magnetic field, a magnetically isotropic material has no preferential direction for its magnetic moment, while a magnetically anisotropic material will align its moment with one of the easy axes. An easy axis is an energetically favourable direction of spontaneous magnetization that is determined by the sources of magnetic anisotropy listed below. The two opposite directions along an easy axis are usually equivalent, and the actual direction of magnetization can be along either of them.

Magnetic anisotropy is a prerequisite for hysteresis in ferromagnets: without it, a ferromagnet is superparamagnetic.

Sources of Magnetic Anisotropy

There are several sources of magnetic anisotropy:

- Magnetocrystalline anisotropy: the atomic structure of a crystal introduces preferential directions for the magnetization.
- Shape anisotropy: when a particle is not perfectly spherical, the demagnetizing field will not be equal for all directions, creating one or more easy axes.
- Magnetoelastic anisotropy: tension may alter magnetic behaviour, leading to magnetic anisotropy.
- Exchange anisotropy: occurs when antiferromagnetic and ferromagnetic materials interact.

Anisotropy Energy of a Single-Domain Magnet

Suppose that a ferromagnet is single-domain in the strictest sense: the magnetization is uniform and rotates in unison. If the magnetic moment is μ and the volume of the particle is V, the magnetization is $\mathrm{M} = \mu / V = M_s(\alpha,\beta,\gamma)$, where M_s is the saturation magnetization and α,β,γ are direction cosines (components of a unit vector) so $\alpha^2+\beta^2+\gamma^2=1$. The energy associated with magnetic anisotropy can depend on the direction cosines in various ways, the most common of which are discussed below.

Uniaxial

A magnetic particle with uniaxial anisotropy has one easy axis. If the easy axis is in the *z* direction, the anisotropy energy can be expressed as one of the forms:

$$E = KV\left(1-\gamma^2\right) = KV\sin^2\theta,$$

where υ is the volume, K the anisotropy constant, and θ the angle between the easy axis and the particle's magnetization. When shape anisotropy is explicitly considered, the symbol $\mathcal{N}$ is often used to indicate the anisotropy constant, instead of K. In the widely used Stoner–Wohlfarth model, the anisotropy is uniaxial.

Triaxial

A magnetic particle with triaxial anisotropy still has a single easy axis, but it also has a hard axis (direction of maximum energy) and an intermediate axis (direction associated with a saddle point in the energy). The coordinates can be chosen so the energy has the form

$$E = K_a V\alpha^2 + K_b V\beta^2.$$

If $K_a > K_b > 0$, the easy axis is the z direction, the intermediate axis is the y direction and the hard axis is the x direction.

Cubic

A magnetic particle with cubic anisotropy has three or four easy axes, depending on the anisotropy parameters. The energy has the form

$$E = KV\left(\alpha^2\beta^2 + \beta^2\gamma^2 + \gamma^2\alpha^2\right).$$

If $K > 0$, the easy axes are the $x, y,$ and z axes. If $K < 0$, there are four easy axes characterized by $x = \pm y = \pm z$.

Iron Oxide Nanoparticles

Iron oxide nanoparticles are iron oxide particles with diameters between about 1 and 100 nanometres. The two main forms are magnetite (Fe_3O_4) and its oxidized form maghemite (γ-Fe_2O_3). They have attracted extensive interest due to their superparamagnetic properties and their potential applications in many fields (although Co and Ni are also highly magnetic materials, they are toxic and easily oxidized). Applications of iron oxide nanoparticles include terabit magnetic storage devices, catalysis, sensors, and high-sensitivity biomolecular magnetic resonance imaging (MRI) for medical diagnosis and therapeutics. These applications require coating of the nanoparticles by agents such as long-chain fatty acids, alkyl-substituted amines and diols.

Structure

Magnetite has an inverse spinel structure with oxygen forming a face-centred cubic crystal system. In magnetite, all tetrahedral sites are occupied by Fe3+and octahedral sites are occupied by both Fe3+and Fe2+. Maghemite differs from magnetite in that all or most of the iron is in the trivalent state (Fe3+) and by the presence of cation vacancies in the octahedral sites. Maghemite has a cubic unit cell in which each cell contains 32 O ions, $21^1/_3$ Fe3+ions and $2^2/_3$ vacancies. The cations are distributed randomly over the 8 tetrahedral and 16 octahedral sites.

Magnetic Properties

Due to its 4 unpaired electrons in 3d shell, an iron atom has a strong magnetic moment. Ions Fe2+ have also 4 unpaired electrons in 3d shell and Fe3+ have 5 unpaired electrons in 3d shell. Therefore, when crystals are formed from iron atoms or ions Fe2+ and Fe3+ they can be in ferromagnetic, antiferromagnetic or ferrimagnetic states.

In the paramagnetic state, the individual atomic magnetic moments are randomly oriented, and the substance has a zero net magnetic moment if there is no magnetic field. These materials have a relative magnetic permeability greater than one and are attracted to magnetic fields. The magnetic moment drops to zero when the applied field is removed. But in a ferromagnetic material, all the atomic moments are aligned even without an external field. A ferrimagnetic material is similar to a ferromagnet but has two different types of atoms with opposing magnetic moments. The material has a magnetic moment because the opposing moments have different strengths. If they have the same magnitude, the crystal is antiferromagnetic and possesses no net magnetic moment.

When an external magnetic field is applied to a ferromagnetic material, the magnetization (*M*) increases with the strength of the magnetic field (*H*) until it approaches saturation. Over some range of fields the magnetization has hysteresis because there is more than one stable magnetic state for each field. Therefore, a remanent magnetization will be present even after removing the external magnetic field.

A single domain magnetic material (ex: magnetic nanoparticles) that has no hysteresis loop is said to be superparamagnetic. The ordering of magnetic moments in ferromagnetic, antiferromagnetic, and ferrimagnetic materials decreases with increasing temperature. Ferromagnetic and ferrimagnetic materials become disordered and lose their magnetization beyond the Curie temperature "T_C" and antiferromagnetic materials lose their magnetization beyond the Néel temperature T_N. Magnetite is ferrimagnetic at room temperature and has a Curie temperature of 850 K. Maghemite is ferrimagnetic at room temperature, unstable at high temperatures, and loses its susceptibility with time. (Its Curie temperature is hard to determine). Both magnetite and maghemite nanoparticles are superparamagnetic at room temperature. This superparamagnetic behaviour of iron oxide nanoparticles can be attributed to their size. When the size gets small enough (<20 nm), thermal fluctuations can change the direction of magnetization of the entire crystal. A material with many such crystals behaves like a paramagnet, except that the moments of entire crystals are fluctuating instead of individual atoms.

Synthesis

The preparation method has a large effect on shape, size distribution, and surface chemistry of the particles. It also determines to a great extent the distribution and type of structural defects or impurities in the particles. All these factors affect magnetic behaviour. Recently, many attempts have been made to develop processes and techniques that would yield 'monodisperse colloids' consisting of nanoparticles uniform in size and shape.

Coprecipitation

By far the most employed method is coprecipitation. This method can be further divided into two types. In the first, ferrous hydroxide suspensions are partially oxidized with different oxidizing agents. For example, spherical magnetite particles of narrow size distribution with mean diameters between 30 and 100 nm can be obtained from a Fe(II) salt, a base and a mild oxidant (nitrate ions). The other method consists in ageing stoichiometric mixtures of ferrous and ferric hydroxides in aqueous media, yielding spherical magnetite particles homogeneous in size. In the second type, the following chemical reaction occurs:

$$2Fe3{+}{+}\ Fe2{+}{+}\ 8OH{+}\text{-}\!\rightarrow Fe_3O_4 + 4H_2O$$

Optimum conditions for this reaction are pH between 8 and 14, Fe3+/Fe2+ratio of 2:1 and a non-oxidizing environment. Being highly susceptibile to oxidation, magnetite (Fe_3O_4) is transformed to maghemite (γFe_2O_3) in the presence of oxygen:

$$2Fe_3O_4 + O_2 \rightarrow 2\gamma Fe_2O_3$$

The size and shape of the nanoparticles can be controlled by adjusting pH, ionic strength, temperature, nature of the salts (perchlorates, chlorides, sulfates, and nitrates), or the Fe(II)/Fe(III) concentration ratio.

Microemulsions

A microemulsion is a stable isotropic dispersion of 2 immiscible liquids consisting of nanosized domains of one or both liquids in the other stabilized by an interfacial film of surface-active molecules. Microemulsions may be categorized further as oil-in-water (o/w) or water-in-oil (w/o), depending on the dispersed and continuous phases.

Water-in-oil is more popular for synthesizing many kinds of nanoparticles. The water and oil are mixed with an amphiphillic surfactant. The surfactant lowers the surface tension between water and oil, making the solution transparent. The water nanodroplets act as nanoreactors for synthesizing nanoparticles. The shape of the water pool is spherical. The size of the nanoparticles will depend on size of the water pool to a great extent. Thus, the size of the spherical nanoparticles can be tailored and tuned by changing the size of the water pool.

High-Temperature Decomposition of Organic Precursors

The decomposition of iron precursors in the presence of hot organic surfactants results in samples with good size control, narrow size distribution (5-12 nm) and good crystallinity; and the nanoparticles are easily dispersed. For biomedical applications like magnetic resonance imaging, magnetic cell separation or magnetorelaxometry, where particle size plays a crucial role, magnetic nanoparticles produced by this method are very useful. Viable iron precursors include Fe(Cup)3,Fe(CO)5, or Fe(acac)3 in organic solvents with surfactant molecules. A combination of Xylenes and Sodium Dodecylbenezensulfonate as a surfactant are used to create nanoreactors for which well dispersed iron(II) and iron (III) salts can react.

Biomedical Applications

Magnetite and maghemite are preferred in biomedicine because they are biocompatible and potentially non-toxic to humans. Iron oxide is easily degradable and therefore useful for in vivo applications. Results from exposure of a human mesothelium cell line and a murine fibroblast cell line to seven industrially important nanoparticles showed a nanoparticle specific cytotoxic mechanism for uncoated iron oxide. Solubility was found to strongly influence the cytotoxic response. Labelling cells (e.g. stem cells, dendritic cells) with iron oxide nanoparticles is an interesting new tool to monitor such labelled cells in real time by magnetic resonance tomography.

5

Electromechanical Actuation of Nanofluids

A Nanofluid is a fluid containing nanometre-sized particles, called nanoparticles. These fluids are engineered colloidal suspensions of nanoparticles in a base fluid. The nanoparticles used in nanofluids are typically made of metals, oxides, carbides, or carbon nanotubes. Common base fluids include water, ethylene glycol and oil.

Nanofluids have novel properties that make them potentially useful in many applications in heat transfer, including microelectronics, fuel cells, pharmaceutical processes, and hybrid-powered engines, engine cooling/vehicle thermal management, domestic refrigerator, chiller, heat exchanger, nuclear reactor coolant, in grinding, machining, in space technology, defense and ships, and in boiler flue gas temperature reduction.

They exhibit enhanced thermal conductivity and the convective heat transfer coefficient compared to the base fluid. Knowledge of the rheological behaviour of nanofluids is found to be very critical in deciding their suitability for convective heat transfer applications.

In analysis such as computational fluid dynamics (CFD), nanofluids can be assumed to be single phase fluids. Classical theory of single phase fluids can be applied, where physical properties of nanofluid is taken as a function of properties of both constituents and their concentrations. An alternative approach simulates nanofluids using a two-component model.

Synthesis of Nanofluids

Nanofluids are supplied by two methods called the one-step and two-step methods. Several liquids including water, ethelene glycol, and oils have been used as base fluids. Nano-materials used so far in nanofluid synthesis include metallic particles, oxide particles, carbon nanotubes, graphene nano-flakes and ceramic particles.

Smart Cooling Nanofluids

Realising the modest thermal conductivity enhancement in conventional nanofluids, a team of researchers at Indira Gandhi Centre for Atomic Research Centre, Kalpakkam developed a new class of magnetically polarizable nanofluids where the thermal conductivity enhancement up to 300% of basefluids is demonstrated. Fatty-acid-capped magnetite nanoparticles of different sizes (3-10 nm) have been synthesized for this purpose. It has been shown that both the thermal and rheological properties of such magnetic nanofluids are tunable by varying the magnetic field strength and orientation with respect to the direction of heat flow. Such response stimuli fluids are reversibly switchable and have applications in miniature devices such as micro- and nano-electromechanical systems. Recently Azizian et al., considered effect of an external magnetic field on the convective heat transfer coefficient of water-based magnetite nanofluid experimentally under laminar flow regime. Up to 300% enhancement obtained at Re=745 and magnetic field gradient of 32.5 mT/mm. The effect of the magnetic field on the pressure drop was not as significant.

Response Stimuli Nanofluids for Sensing Applications

Researchers have invented a nanofluid based ultrasensitive optical sensor that changes its colour on exposure to extremely low concentrations of toxic cations. The sensor is useful in detecting minute traces of cations in industrial and environmental samples. Existing techniques for monitoring cations levels in industrial and environmental samples are expensive, complex and time-consuming. The sensor is designed with a magnetic nanofluid that consists of nanodroplets with magnetic grains suspended in water. At a fixed magnetic field, a light source illuminates the nanofluid where the colour of the nanofluid changes depending on the cation concentration. This colour change occurs within a second after exposure to cations, much faster than other existing cation sensing methods.

Such response stimulus nanofluids are also used to detect and image defects in ferromagnetic components. The photonic eye, as it has been called, is based on a magnetically polarizable nano-emulsion that changes colour when it comes into contact with a defective region in a sample. The device might be used to monitor structures such as rail tracks and pipelines.

Applications

Nanofluids are primarily used as coolant in heat transfer equipment such as heat exchangers, electronic cooling system(such as flat plate) and radiators. Heat transfer over flat plate has been analyzed by a lot of researchers. Graphene based nanofluid has been found to enhance Polymerase chain reaction efficiency. Nanofluids in solar collectors is another application where nanofluids are employed for their tunable optical properties.

Colloid

A colloid is a substance microscopically dispersed throughout another substance.

The dispersed-phase particles have a diameter of between approximately 1 and 1000 nanometres. Such particles are normally easily visible in an optical microscope, although at the smaller size range ($r<250$nm), an ultramicroscope or an electron microscope may be required. Homogeneous mixtures with a dispersed phase in this size range may be called *colloidal aerosols*, *colloidal emulsions*, *colloidal foams*, *colloidal dispersions*, or *hydrosols*. The dispersed-phase particles or droplets are affected largely by the surface chemistry present in the colloid.

Some colloids are translucent because of the Tyndall effect, which is the scattering of light by particles in the colloid. Other colloids may be opaque or have a slight colour.

Colloidal solutions (also called colloidal suspensions) are the subject of interface and colloid science. This field of study was introduced in 1861 by Scottish scientist Thomas Graham.

Classification

Because the size of the dispersed phase may be difficult to measure, and because colloids have the appearance of solutions, colloids are sometimes identified and characterized by their physico-chemical and transport properties. For example, if a colloid consists of a solid phase dispersed in a liquid, the solid particles will not diffuse through a membrane, whereas with a true solution the dissolved ions or molecules will diffuse through a membrane. Because of the size exclusion, the colloidal particles are unable to pass through the pores of an ultrafiltration membrane with a size smaller than their own dimension.

The smaller the size of the pore of the ultrafiltration membrane, the lower the concentration of the dispersed colloidal particles remaining in the ultrafiltered liquid. The exact value of the concentration of a truly dissolved species will thus depend on the experimental conditions applied to separate it from the colloidal particles also dispersed in the liquid. This is particularly important for solubility studies of readily hydrolysed species such as Al, Eu, Am, Cm, or organic matter complexing these species. Colloids can be classified as follows:

Medium/phase		***Dispersed phase***		
		Gas	***Liquid***	***Solid***
Dispersion medium	Gas	None All gases are miscible and thus do not form colloids	Liquid aerosol Examples: fog, hair sprays	Solid aerosol Examples: smoke, ice cloud, air particulates
	Liquid	Foam Example: whipped cream, shaving cream	Emulsion Examples: milk, mayonnaise, hand cream	Sol Examples: pigmented ink, blood
	Solid	Solid foam Examples: aerogel, styrofoam, pumice	Gel Examples: agar, gelatin, jelly	Solid sol Example: cranberry glass

Based on the nature of interaction between the dispersed phase and the dispersion medium, colloids can be classified as: Hydrophilic colloids: These are water-loving colloids. The colloid particles are attracted toward water. They are also called reversible sols. Hydrophobic colloids: These are opposite in nature to hydrophilic colloids. The colloid particles are repelled by water. They are also called irreversible sols.

In some cases, a colloid can be considered a homogeneous mixture. This is because the distinction between "dissolved" and "particulate" matter can be sometimes a matter of approach, which affects whether or not it is homogeneous or heterogeneous.

Hydrocolloids

A hydrocolloid is defined as a colloid system wherein the colloid particles are hydrophilic polymers dispersed in water. A hydrocolloid has colloid particles spread throughout water, and depending on the quantity of water available that can take place in different states, e.g., gel or sol (liquid). Hydrocolloids can be either irreversible (single-state) or reversible. For example, agar, a reversible hydrocolloid of seaweed extract, can exist in a gel and solid state, and alternate between states with the addition or elimination of heat.

Many hydrocolloids are derived from natural sources. For example, agar-agar and carrageenan are extracted from seaweed, gelatin is produced by hydrolysis of proteins of bovine and fish origins, and pectin is extracted from citrus peel and apple pomace.

Gelatin desserts like jelly or Jell-O are made from gelatin powder, another effective hydrocolloid. Hydrocolloids are employed in food mainly to influence texture or viscosity (e.g., a sauce). Hydrocolloid-based medical dressings are used for skin and wound treatment.

Other main hydrocolloids are xanthan gum, gum arabic, guar gum, locust bean gum, cellulose derivatives as carboxymethyl cellulose, alginate and starch.

Based on the nature of interaction between the dispersed phase and the dispersion medium, colloids can be classified as: Hydrophilic colloids: These are water-loving colloids. The colloid particles are attracted toward water. They are also called reversible sols. Hydrophobic colloids: These are opposite in nature to hydrophilic colloids. The colloid particles are repelled by water. They are also called irreversible sols.

Interaction between Particles

The following forces play an important role in the interaction of colloid particles:

- Excluded volume repulsion: This refers to the impossibility of any overlap between hard particles.
- Electrostatic interaction: Colloidal particles often carry an electrical charge and therefore attract or repel each other. The charge of both the continuous and the dispersed phase, as well as the mobility of the phases are factors affecting this interaction.
- van der Waals forces: This is due to interaction between two dipoles that are either permanent or induced. Even if the particles do not have a permanent dipole, fluctuations of the electron density gives rise to a temporary dipole in a particle. This temporary dipole induces a dipole in particles nearby. The temporary dipole and the induced dipoles are then attracted to each other. This is known as van der Waals force, and is always present (unless the refractive indexes of the dispersed and continuous phases are matched), is short-range, and is attractive.
- Entropic forces: According to the second law of thermodynamics, a system progresses to a state in which entropy is maximized. This can result in effective forces even between hard spheres.
- Steric forces between polymer-covered surfaces or in solutions containing non-adsorbing polymer can modulate interparticle forces, producing an additional steric repulsive force (which is predominantly entropic in origin) or an attractive depletion force between them. Such an effect is specifically searched for with

tailor-made superplasticizers developed to increase the workability of concrete and to reduce its water content.

Preparation

There are two principal ways of preparation of colloids:

- Dispersion of large particles or droplets to the colloidal dimensions by milling, spraying, or application of shear (e.g., shaking, mixing, or high shear mixing).
- Condensation of small dissolved molecules into larger colloidal particles by precipitation, condensation, or redox reactions. Such processes are used in the preparation of colloidal silica or gold.

Stabilization (Peptization)

The stability of a colloidal system is defined by particles remaining suspended in solution at equilibrium.

Stability is hindered by aggregation and sedimentation phenomena, which are driven by the colloids tendency to reduce surface energy. Reducing the interfacial tension will stabilize the colloidal system by reducing this driving force.

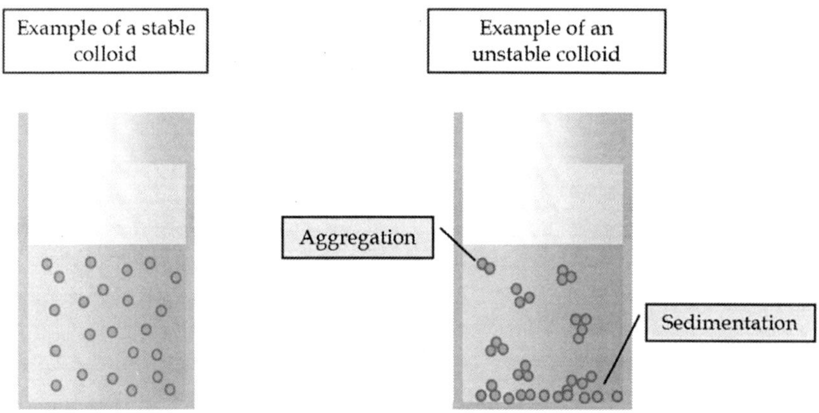

Figure: *Examples of a stable and of an unstable colloidal dispersion.*

Aggregation is due to the sum of the interaction forces between particles. If attractive forces (such as van der Waals forces) prevail over the repulsive ones (such as the electrostatic ones) particles aggregate in clusters.

Electrostatic stabilization and steric stabilization are the two main mechanisms for stabilization against aggregation.

- Electrostatic stabilization is based on the mutual repulsion of like electrical charges. In general, different phases have different charge affinities, so that an electrical double layer forms at any interface. Small particle sizes lead to enormous surface areas, and this effect is greatly amplified in colloids. In a stable colloid, mass of a dispersed phase is so low that its buoyancy or kinetic energy is too weak to overcome the electrostatic repulsion between charged layers of the dispersing phase.

- Steric stabilization consists in covering the particles in polymers which prevents the particle to get close in the range of attractive forces.

A combination of the two mechanisms is also possible (electrosteric stabilization). All the above mentioned mechanisms for minimizing particle aggregation rely on the enhancement of the repulsive interaction forces. Electrostatic and steric stabilization do not directly address the sedimentation/floating problem.

Particle sedimentation (and also floating, although this phenomenon is less common) arises from a difference in the density of the dispersed and of the continuous phase. The higher the difference in densities, the faster the particle settling.

- The gel network stabilization represents the principal way to produce colloids stable to both aggregation and sedimentation.

The method consists in adding to the colloidal suspension a polymer able to form a gel network and characterized by shear thinning properties. Examples of such substances are xanthan and guar gum.

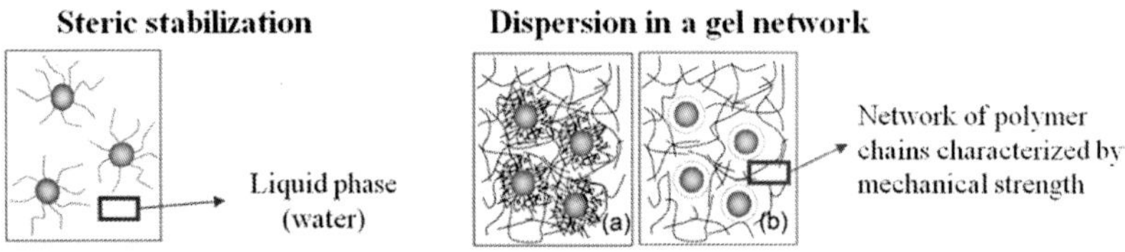

Figure: *Steric and Gel network stabilization.*

Particle settling is hindered by the stiffness of the polymeric matrix where particles are trapped. In addition, the long polymeric chains can provide a steric or electrosteric stabilization to dispersed particles. The rheological shear thinning properties find beneficial in the preparation of the suspensions and in their use, as the reduced viscosity at high shear rates facilitates deagglomeration, mixing and in general the flow of the suspensions.

Destabilization (Flocculation)

Unstable colloidal dispersions can form flocs as the particles aggregate due to interparticle attractions. In this way photonic glasses can be grown. This can be accomplished by a number of different methods:

- Removal of the electrostatic barrier that prevents aggregation of the particles. This can be accomplished by the addition of salt to a suspension or changing the pH of a suspension to effectively neutralize or "screen" the surface charge of the particles in suspension. This removes the repulsive forces that keep colloidal particles separate and allows for coagulation due to van der Waals forces.
- Addition of a charged polymer flocculant. Polymer flocculants can bridge individual colloidal particles by attractive electrostatic interactions. For example, negatively-charged colloidal silica or clay particles can be flocculated by the addition of a positively-charged polymer.
- Addition of non-adsorbed polymers called depletants that cause aggregation due to entropic effects.
- Physical deformation of the particle (e.g., stretching) may increase the van der Waals forces more than stabilization forces (such as electrostatic), resulting coagulation of colloids at certain orientations.

Unstable colloidal suspensions of low-volume fraction form clustered liquid suspensions, wherein individual clusters of particles fall to the bottom of the suspension (or float to the top if the particles are less dense than the suspending medium) once the clusters are of sufficient size for the Brownian forces that work to keep the particles in suspension to be overcome by gravitational forces.

However, colloidal suspensions of higher-volume fraction form colloidal gels with viscoelastic properties. Viscoelastic colloidal gels, such as bentonite and toothpaste, flow like liquids under shear, but maintain their shape when shear is removed. It is for this reason that toothpaste can be squeezed from a toothpaste tube, but stays on the toothbrush after it is applied.

Monitoring Stabilitye

Multiple light scattering coupled with vertical scanning is the most widely used technique to monitor the dispersion state of a product, hence identifying and quantifying destabilisation phenomena. It works on concentrated dispersions without dilution. When light is sent through the sample, it is backscattered by the particles / droplets. The backscattering intensity is directly proportional to the size and volume fraction of the dispersed phase. Therefore, local changes in concentration (*e.g.* Creaming and Sedimentation) and global changes in size (*e.g.* flocculation, coalescence) are detected and monitored.

Accelerating Methods for Shelf Life Prediction

The kinetic process of destabilisation can be rather long (up to several months or even years for some products) and it is often required for the formulator to use further accelerating methods in order to reach reasonable development time for new product design. Thermal methods are the most commonly used and consists in increasing temperature to accelerate destabilisation (below critical temperatures of phase inversion or chemical degradation).

Temperature affects not only the viscosity, but also interfacial tension in the case of non-ionic surfactants or more generally interactions forces inside the system. Storing a dispersion at high temperatures enables to simulate real life conditions for a product (e.g. tube of sunscreen cream in a car in the summer), but also to accelerate destabilisation processes up to 200 times. Mechanical acceleration including vibration, centrifugation and agitation are sometimes used. They subject the product to different forces that pushes the particles / droplets against one another, hence helping in the film drainage. However, some emulsions would never coalesce in normal gravity, while they do under artificial gravity. Moreover segregation of different populations of particles have been highlighted when using centrifugation and vibration.

As a Model System for Atoms

In physics, colloids are an interesting model system for atoms. Micrometre-scale colloidal particles are large enough to be observed by optical techniques such as confocal microscopy. Many of the forces that govern the structure and behaviour of matter, such as excluded volume interactions or electrostatic forces, govern the structure and behaviour of colloidal suspensions. For example, the same techniques used to model ideal gases can be applied to model the behaviour of a hard sphere colloidal suspension. In addition, phase transitions in colloidal suspensions can be studied in real time using optical techniques, and are analogous to phase transitions in liquids. In many interesting cases optical fluidity is used to control colloid suspensions.

Colloidal Crystal

A colloidal crystal is an ordered array of colloid particles, analogous to a standard crystal whose repeating subunits are atoms or molecules. A natural example of this phenomenon can be found in the gem opal, where spheres of silica assume a close-packed locally periodic structure under moderate compression. Bulk properties of a colloidal crystal depend on composition, particle size, packing arrangement, and degree of regularity. Applications include photonics, materials processing, and the study of self-assembly and phase transitions.

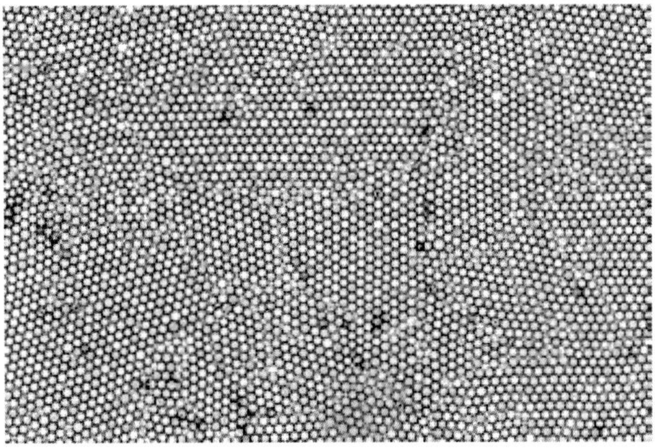

Figure: *A collection of small 2D colloidal crystals with grain boundaries between them. Spherical glass particles (10 μm diameter) in water.*

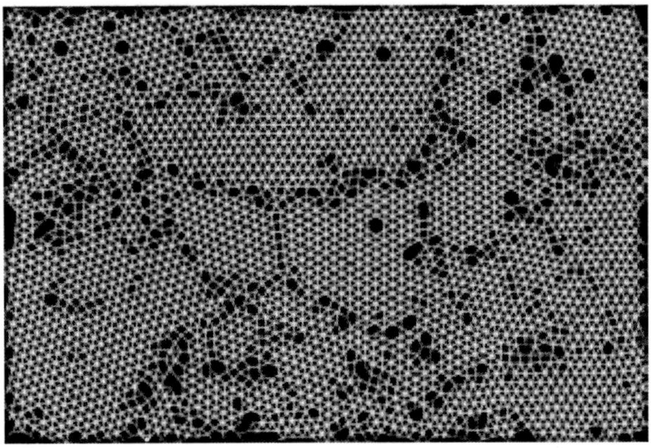

Figure: *The connectivity of the crystals in the colloidal crystals above. Connections in white indicate that particle has six equally spaced neighbours and therefore forms part of a crystalline domain.*

The origins of colloidal crystals go back to the mechanical properties of bentonite sols, and the optical properties of Schiller layers in iron oxide sols. The properties are supposed to be due to the ordering of monodisperse inorganic particles. Monodisperse colloids, capable of forming long-range ordered arrays, existing in nature. The discovery by W.M. Stanley of the crystalline forms of the tobacco and tomato viruses provided examples of this. Using X-ray diffraction methods, it was subsequently determined that when concentrated by centrifuging from dilute water suspensions, these virus particles often organised themselves into highly ordered arrays.

Rod-shaped particles in the tobacco mosaic virus could form a two-dimensional triangular lattice, while a body-centred cubic structure was formed from the almost spherical particles in the tomato Bushy Stunt Virus. In 1957, a letter describing the discovery of "*A Crystallizable Insect Virus*" was published in the journal *Nature*. Known as the Tipula Iridiscent Virus, from both square and triangular arrays occurring on crystal faces, the authors deduced the face-centred cubic close-packing of virus particles. This type of ordered array has also been observed in cell suspensions, where the symmetry is well adapted to the mode of reproduction of the organism. The limited content of genetic material places a restriction on the size of the protein to be coded by it. The use of a large number of the same proteins to build a protective shell is consistent with the limited length of RNA or DNA content.

It has been known for many years that, due to repulsive Coulombic interactions, electrically charged macromolecules in an aqueous environment can exhibit long-range crystal-like correlations with interparticle separation distances often being considerably greater than the individual particle diameter. In all of the cases in nature, the same iridescence is caused by the diffraction and constructive interference of visible lightwaves which falls under Bragg's law.

Because of the rarity and pathological properties, neither opal nor any of the organic viruses have been very popular in scientific laboratories. The number of experiments exploring the physics and chemistry of these "colloidal crystals" has emerged as a result of the simple methods which have evolved in 20 years for preparing synthetic monodisperse colloids, both polymer and mineral, and, through various mechanisms, implementing and preserving their long-range order formation.

Trends

Colloidal crystals are receiving increased attention, largely due to their mechanisms of ordering and self-assembly, cooperative motion, structures similar to those observed in condensed matter by both liquids and solids, and structural phase transitions. Phase equilibrium has been considered within the context of their physical similarities, with appropriate scaling, to elastic solids. Observations of the interparticle separation distance has shown a decrease on ordering. This led to a re-evaluation of Langmuir's beliefs about the existence of a long-range attractive component in the interparticle potential.

Colloidal crystals have found application in optics as photonic crystals. Photonics is the science of generating, controlling, and detecting photons (packets of light), particularly in the visible and near Infrared, but also extending to the Ultraviolet, Infrared and far IR portions of the electromagnetic spectrum. The science of photonics includes the emission, transmission, amplification, detection, modulation, and switching of lightwaves over a broad range of frequencies and wavelengths. Photonic devices include electro-optic components such as lasers (Light Amplification by Stimulated Emission of Radiation) and optical fibre. Applications include telecommunications, information processing, illumination, spectroscopy, holography, medicine (surgery, vision correction, endoscopy), military (guided missile) technology, agriculture and robotics.

Polycrystalline colloidal structures have been identified as the basic elements of submicrometre colloidal materials science. Molecular self-assembly has been observed in various biological systems and underlies the formation of a wide variety of complex biological structures. This includes an emerging class of mechanically superior biomaterials based on microstructure features and designs found in nature.

The principal mechanical characteristics and structures of biological ceramics, polymer composites, elastomers, and cellular materials are being re-evaluated, with an emphasis on bioinspired materials and structures. Traditional approaches focus on design methods of biological materials using conventional synthetic materials. The uses have been identified in the synthesis of bioinspired materials through processes that are characteristic of biological systems in nature. This includes the nanoscale self-assembly of the components and the development of hierarchical structures.

Bulk Crystals

Aggregation: Aggregation in colloidal dispersions (or stable suspensions) has been characterized by the degree of interparticle attraction. For attractions strong relative to the thermal energy (given by kT), Brownian motion produces irreversibly flocculated structures with growth rates limited by the rate of particle diffusion. This leads to a description using such parameters as the degree of branching, ramification or fractal dimensionality. A reversible growth model has been constructed by modifying the cluster-cluster aggregation model with a finite inter-particle attraction energy.

In systems where forces of attraction forces are buffered to some degree, a balance of forces leads to an equilibrium phase separation, that is particles coexist with equal chemical potential in two distinct structural phases. The role of the ordered phase as an elastic colloidal solid has been evidenced by the elastic (or reversible) deformation due to the force of gravity. This deformation can be quantified by the distortion of the lattice parameter, or inter-particle spacing.

Viscoelasticity

Periodic ordered lattices behave as linear viscoelastic solids when subjected to small amplitude mechanical deformations. Okano's group experimentally correlated the shear modulus to the frequency of standing shear modes using mechanical resonance techniques in the ultrasonic range (40 to 70 kHz). In oscillatory experiments at lower frequencies (< 40 Hz), the fundamental mode of vibration as well as several higher frequency partial overtones (or harmonics) have been observed. Structurally, most systems exhibit a clear instability toward the formation of periodic domains of relatively short-range order Above a critical amplitude of oscillation, plastic deformation is the primary mode of structural rearrangement.

Phase Transitions

Equilibrium phase transitions (e.g. order/disorder), an equation of state, and the kinetics of colloidal crystallization have all been actively studied, leading to the development of several methods to control the self-assembly of the colloidal particles. Examples include colloidal epitaxy and space-based reduced-gravity techniques, as well as the use of temperature gradients to define a density gradient. This is somewhat counterintuitive as temperature does not play a role in determining the hard-sphere phase diagram. However, hard-sphere single crystals (size 3 mm) have been obtained from a sample in a concentration regime that would remain in the liquid state in the absence of a temperature gradient.

Phonon Dispersion

Using a single colloidal crystal, phonon dispersion of the normal modes of vibration modes were investigated using photon correlation spectroscopy, or dynamic light scattering. This technique relies on the relaxation or decay of concentration (or density) fluctuations. These are often associated with longitudinal modes in the acoustic range. A distinctive increase in the sound wave velocity (and thus the elastic modulus) by a factor of 2.5 has been observed at the structural transition from colloidal liquid to colloidal solid, or point of ordering.

Kossel Lines

Using a single body-centred cubic colloidal crystal, the occurrence of Kossel lines in diffraction patterns were used to monitor the initial nucleation and subsequent motion caused distortion of the crystal. Continuous or homogeneous deformations occurring beyond the elastic limit produce a 'flowing crystal', where the nucleation site density increases significantly with increasing particle concentration. Lattice dynamics have been investigated for longitudinal as well as transverse modes. The same technique was used to evaluate the crystallization process near the edge of a glass tube. The former might be considered analogous to a homogeneous nucleation event—whereas the latter would clearly be considered a heterogeneous nucleation event, being catalyzed by the surface of the glass tube.

Growth Rates

Small-angle laser light scattering has provided information about spatial density fluctuations or the shape of growing crystal grains. In addition, confocal laser scanning microscopy has been used to observe crystal growth near a glass surface. Electro-optic shear waves have been induced by an ac pulse, and monitored by reflection spectroscopy as well as light scattering. Kinetics of colloidal crystallization have been measured quantitatively, with nucleation rates being depending on the suspension concentration. Similarly, crystal growth rates have been shown to decrease linearly with increasing reciprocal concentration.

Microgravity

Experiments performed in microgravity on the Space Shuttle Columbia suggest that the typical face-centred cubic structure may be induced by gravitational stresses. Crystals tend to exhibit the hcp structure alone (random stacking of hexagonally close-packed crystal planes), in contrast with a mixture of (rhcp) and face-centred cubic packing when allowed sufficient time to reach mechanical equilibrium under gravitational forces on Earth. Glassy (disordered or amorphous) colloidal samples have become fully crystallized in microgravity in less than two weeks.

Thin Films

Two-dimensional (thin film) semi-ordered lattices have been studied using an optical microscope, as well as those collected at electrode surfaces. Digital video microscopy has revealed the existence of an equilibrium hexatic phase as well as a strongly first-order liquid-to-hexatic and hexatic-to-solid phase transition. These observations are in agreement with the explanation that melting might proceed via the unbinding of pairs of lattice dislocations.

Long-Range Order

Long-range order has been observed in thin films of colloidal liquids under oil—with the faceted edge of an emerging single crystal in alignment with the diffuse streaking pattern in the liquid phase. Structural defects have been directly observed in the ordered solid phase as well as at the interface of the solid and liquid phases. Mobile lattice defects have been observed via Bragg reflections, due to the modulation of the light waves in the strain field of the defect and its stored elastic strain energy.

Mobile Lattice Defects

All of the experiments have led to at least one common conclusion: colloidal crystals may indeed mimic their atomic counterparts on appropriate scales of length (spatial) and time (temporal). Defects have been reported to flash by in the blink of an eye in thin films of colloidal crystals under oil using a simple optical microscope. But quantitatively measuring the rate of its propagation provides an entirely different challenge, which has been measured at somewhere near the speed of sound.

Non-Spherical Colloid Based Crystal

Crystalline thin-films from non-spherical colloids were produced using convective assembly techniques. Colloid shapes included dumbbell, hemisphere, disc, and sphero-cylinder shapes. Both purely crystalline and plastic crystal phases could be produced, depending on the aspect ratio of the colloidal particle. The particles were crystallized both as 2D (i.e., monolayer) and 3D (i.e., multilayer) structures. The observed lattice and particle orientations experimentally confirmed a body of theoretical work on the condensed phases of non-spherical objects.

Applications

Photonics: Technologically, colloidal crystals have found application in the world of optics as photonic band gap (PBG) materials (or photonic crystals). Synthetic opals as well as inverse opal configurations are being formed either by natural sedimentation or applied forces, both achieving similar results: long-range ordered structures which provide a natural diffraction grating for lightwaves of wavelength comparable to the particle size.

Novel PBG materials are being formed from opal-semiconductor-polymer composites, typically utilising the ordered lattice to create an ordered array of holes (or pores) which is left behind after removal or decomposition of the original particles. Residual hollow honeycomb structures provide a relative index of refraction (ratio of matrix to air) sufficient for selective filters. Variable index liquids or liquid crystals injected into the network alter the ratio and band gap.

Such frequency-sensitive devices may be ideal for optical switching and frequency selective filters in the ultraviolet, visible, or infrared portions of the spectrum, as well as higher efficiency antennae at microwave and millimetre wave frequencies.

Self-Assembly

Self-assembly is the most common term in use in the modern scientific community to describe the spontaneous aggregation of particles (atoms, molecules, colloids, micelles, etc.) without the influence of any external forces. Large groups of such particles are known to assemble themselves into thermodynamically stable, structurally well-defined arrays, quite reminiscent of one of the 7 crystal systems found in metallurgy and mineralogy (e.g. face-centred cubic, body-centred cubic, etc.). The fundamental difference in equilibrium structure is in the spatial scale of the unit cell (or lattice parameter) in each particular case.

Molecular self-assembly is found widely in biological systems and provides the basis of a wide variety of complex biological structures. This includes an emerging class of mechanically superior biomaterials based on microstructural features and designs found in nature. Thus, self-assembly is also emerging as a new strategy in chemical synthesis and nanotechnology. Molecular crystals, liquid crystals, colloids, micelles, emulsions, phase-separated polymers, thin films and self-assembled monolayers all represent examples of the types of highly ordered structures which are obtained using these techniques. The distinguishing feature of these methods is self-organisation.

Colloid in Biology

In the early 20th century, before enzymology was well understood, colloids were thought to be the key to the operation of enzymes; i.e., the addition of small quantities of an enzyme to a quantity of water would, in some fashion yet to be specified, subtly alter the properties of the water so that it would break down the enzyme's specific substrate, such as a solution of ATPase breaking down ATP. Furthermore, life itself was explainable in terms of the aggregate properties of all the colloidal substances that make up an organism. As more detailed knowledge of biology and biochemistry developed, the colloidal theory was replaced by the macromolecular theory, which explains an enzyme as a collection of identical huge molecules that act as very tiny machines, freely moving about between the water molecules of the solution and individually operating on the substrate, no more mysterious than a factory full of machinery. The properties of the water in the solution are not altered, other than the simple osmotic changes that would be caused by the presence of any solute. In humans, both the thyroid gland and the intermediate lobe (*pars intermedia*) of the pituitary gland contain colloid follicles.

Colloid in the Environment

Colloidal particles can also serve as transport vector of diverse contaminants in the surface water (sea water, lakes, rivers, fresh water bodies) and in underground water circulating in fissured rocks (limestone, sandstone, granite, ...). Radionuclides and heavy metals easily sorb onto colloids suspended in water. Various types of colloids are recognised: inorganic colloids (clay particles, silicates, iron oxy-hydroxides, ...), organic colloids (humic and fulvic substances). When heavy metals or radionuclides form their own pure colloids, the term "*Eigencolloid*" is used to designate pure phases, e.g., $Tc(OH)_4$, $U(OH)_4$, $Am(OH)_3$. Colloids have been suspected for the long-range transport of plutonium on the Nevada Nuclear Test Site. They have been the subject of detailed studies for many years. However, the mobility of inorganic colloids is very low in compacted bentonites and in deep clay formations because of the process of ultrafiltration occurring in dense clay membrane. The question is less clear for small organic colloids often mixed in porewater with truly dissolved organic molecules.

Intravenous Therapy

Colloid solutions used in intravenous therapy belong to a major group of volume expanders, and can be used for intravenous fluid replacement. Colloids preserve a high colloid osmotic pressure in the blood, and therefore, they should theoretically preferentially increases the intravascular volume, whereas other types of volume expanders called crystalloids also increases the interstitial volume and intracellular volume. However, there is still controversy to the actual difference in efficacy by this difference. Another difference is that crystalloids generally are much cheaper than colloids. Recently, however, it has been determined that the use of colloids was bolstered by faked research studies.

Nanoparticle Applications for Environmental Control and Remediation

Nanotechnology is an emerging field that covers a wide range of technologies which are presently under development in nanoscale. It plays a major role in the development of innovative methods to produce new products, to substitute existing production equipment and to reformulate new materials and chemicals with improved performance resulting in less consumption of energy and materials and reduced harm to the environment as well as environmental remediation. Although, reduced consumption of energy and materials benefits the environment, nanotechnology will give possibilities to remediate problems associated with the existing processes in a more sustainable way. Environmental applications of nanotechnology address the development of solutions to the existing environmental problems, preventive measures for future problems resulting from the interactions of energy and materials with the environment, and any possible risks that may be posed by nanotechnology itself.

Nanotechnology is a field of applied science, focused on the design, synthesis, characterization and application of materials and devices on the nanoscale.

This branch of knowledge is a sub-classification of technology in colloidal science, biology, physics, chemistry and other scientific fields and involves the study of phenomena and manipulation of materials in the nanoscale [1-3]. This results in materials and systems that often exhibit novel and significantly changing physical, chemical and biological properties due to their size and structure [3]. Also, a unique aspect of nanotechnology is the "vastly increased ratio of surface area to volume", present in many nanoscale materials, which opens new possibilities in surface-based sciences.

Similar to nanotechnology's success in consumer products and other sectors, nanoscale materials have the potential to improve the environment, both through direct applications of those materials to detect, prevent, and remove pollutants, as well as indirectly by using nanotechnology to design cleaner industrial processes and create environmentally responsible products. For example, iron nanoparticles can remove contaminants from soil and ground water; and nanosized sensors hold promise for improved detection and tracking of contaminants.

Behaviour of materials at nanoscale is not necessarily predictable from what we know at macroscale. At the nanoscale, often highly desirable, properties are created due to size confinement, dominance of interfacial phenomena, and quantum effects. These new and unique properties of nanostructured materials, nanoparticles, and other related nanotechnologies lead to improved catalysts, tunable photoactivity, increased strength, and many other interesting characteristics.

As the exciting field of nanotechnology develops, the broader environmental impacts of nanotechnology will also need to be considered. Such considerations might include: the environmental implications of the cost, size and availability of advanced technological devices; models to determine potential benefits of reduction or prevention of pollutants from environmental sources; potential new directions in environmental science due to advanced sensors; effects of rapid advances in health care and health management as related to the environment; impact of artificial nanoparticles in the atmosphere; and impacts from the development of nanomachines.

Research is needed using nanoscale science and technology to identify opportunities and applications to environmental problems, and to evaluate the potential environmental impacts of nanotechnology. Also, approaches are needed to offer new capabilities for preventing or treating highly toxic or persistent pollutants, which would result in the more effective monitoring of pollutants or their impact in the ways not currently possible.

Early application of nanotechnology is remediation using nanoscale iron particles. Zero-valent iron nanoparticles are deployed in situ to remediate soil and water contaminated with chlorinated compounds and heavy metals.

Among the many applications of nanotechnology that have environmental implications, remediation of contaminated groundwater using nanoparticles containing zero-valent iron is one of the most prominent examples of a rapidly emerging technology with considerable potential benefits. There are, however, many uncertainties regarding the fundamental features of this technology, which have made it difficult to engineer applications for optimal performance or to assess the risk to human or ecological health. This important aspect of nanoparticles needs extensive considerations as well.

One of the main environmental applications of nanotechnology is in the water sector. As freshwater sources become increasingly scarce due to overconsumption and contamination, scientists have begun to consider seawater as another source for drinking water. The majority of the world's water supply has too much salt for human consumption and desalination is an option but expensive method for removing the salt to create new sources of drinking water. Carbon nanotube membranes have the potential to reduce desalination costs. Similarly, nanofilters could be used to remediate or clean up ground water or surface water contaminated with chemicals and hazardous substances. Finally, nanosensors could be developed to detect waterborne contaminants.

Air pollution is another potential area where nanotechnology has great promise. Filtration techniques similar to the water purification methods described above could be used in buildings to purify indoor air volumes. Nanofilters could be applied to automobile tailpipes and factory smokestacks to separate out contaminants and prevent them from entering the atmosphere. Finally, nanosensors could be developed to detect toxic gas leaks at extremely low concentrations. Overall, there is a multitude of promising environmental applications for nanotechnology. Much of the current research is focused on energy and water technologies.

Environmental remediation includes the degradation, sequestration, or other related approaches that result in reduced risks to human and environmental receptors posed by chemical and radiological contaminants. The benefits, which arise from the application of nanomaterials for remediation, would be more rapid or cost-effective cleanup of wastes.

Cost-effective remediation techniques pose a major challenge in the development of adequate remediation methods that protect the environment. Substances of significant concern in remediation of soils, sediment, and groundwater include heavy metals (e.g., mercury, lead, cadmium) and organic compounds (e.g., benzene, chlorinated solvents, creosote, and toluene). Specific control and design of materials at the molecular level may impart increased affinity, capacity, and selectivity for pollutants. Minimizing quantities and exposure of hazardous wastes to the air and water and providing safe drinking water are among the environmental protection agencies' goals. In this regards, nanotechnology could play a key role in pollution prevention technologies.

Nano-Materials and Their Environmental Applications

Titanium Dioxide (TiO2) Based Nanoparticles

Titanium dioxide (TiO2) is one of the popular materials used in various applications because of its semiconducting, photocatalytic, energy converting, electronic and gas sensing properties. Titanium dioxide crystals are present in three different polymorphs in nature that in the order of their abundance, are Rutile, Anatase and Brookite.

Many researchers are focused on TiO2 nanoparticle and its application as a photocatalyst in water treatment. Nanoparticles that are activated by light, such as the large band-gap semiconductors titanium dioxide (TiO2) and zinc oxide (ZnO), are frequently studied for their ability to remove organic contaminants from various media. These nanoparticles have the advantages of readily available, inexpensive, and low toxicity. The semiconducting property of TiO2 is necessary for the removal of different organic pollutants through excitation of TiO2 semiconductor with a light energy greater than its band gap, which could generate electron hole pairs. These may be exploited in different reduction processes at the semiconductor/solution interface.

A semiconductor can adopt with donor atoms that provide electrons for the conduction band where they can carry a current. These materials can also adopt with acceptor atoms that take electrons from the valence band and leave behind some positive charges (holes). The energy levels of these donors and acceptors fall into the energy gap. The most affecting properties of semiconducting nanoparticles are distinguished changes in their optical properties compared to those of bulk materials. In addition, there is a significant shift in optical absorption spectra toward the blue shift (shorter wavelengths) as the particle size is reduced.

Stathatos et al. used reverse micelle technique to produce TiO2 nanoparticles and deposited them as thin films. They deposited TiO2 mesoporous films on glass slides by dip-coating in reverse micellar gels containing titanium isopropoxide. The films demonstrated high capacity in adsorption of several dyes from aqueous and alcoholic solutions. It had also a rapid degradation of the adsorbed dyes when the coloured films were exposured to the visible light.

It is known that, the semiconducting properties of TiO2 materials is responsible for the removal of various organic pollutants, but the rapid recombination of photo-generated electron hole pairs and the non-selectivity of the system are the main problems that limit the application of photocatalysis processes. It was suggested that, replacing adsorbed solvent molecules and ions by chelating agents, i.e. surface modification, changes the energetic situation of surface states and considerably alters the chemistry, which is taking place at the surface of titanium dioxide (TiO2).

Phenol is one of the toxic materials found in municipal and waste waters. Synthesized titanium dioxide nanoparticles of both Anatase and Rutile forms were used for wet oxidation of phenols by hydrothermal treatment of microemulsions and their photocatalytic activity. Such treatment has the advantage that the size of particles is affected by the ratio of surfactant to water. Size of water droplets in the reverse microemulsions is found to be almost the same as that of formed particles.

Fine TiO2 particles have shown better efficiency than the immobilized catalysts, but complete separation and recycling of fine particles (less than 0.5μm) from the treated water, are very expensive.

Therefore, from the economics point of view, this method is not suitable for the industrial-scale. This problem was solved by fixing the carbon-black-modified nano-TiO2 (CB-TiO2) on aluminum sheet as a support. The photocatalytic activity of CB-TiO2 thin films was observed to be 1.5 times greater than that of TiO2 thin films in the degradation of reactive Brilliant Red X-3B. Core $SrFe_{12}O19$ nanoparticles and TiO2 nanocrystals were also synthesized as the magnetic photocatalytic particles. This system recovers the photocatalyst particles and protects them from the treated water stream by applying an external magnetic field. In the presence of natural organic matters in water, many problems may happen since they can occupy the catalytically active surface sites and lead to much lower decomposition efficiency. One useful method for overcoming the mentioned problem is the combination of adsorption and oxidative destruction technique.

Bimetallic Nanoparticles

Destruction of halogenated organic compound (HOCs) by zero-valent iron represents one of the latest innovative technologies for environmental remediation. Laboratory investigation in the past few years indicated that granular iron could degrade many HOCs, such as chlorinated aliphatics, chlorinated aromatics and polychlorinated biphenyls. Wang and Zhang stated that the implementation of zero-valent iron technique would encounter challenges such as:

Production and accumulation of chlorinated by-products due to the low reactivity of iron powders toward lightly chlorinated hydrocarbons. For example, reduction of tetrachloroethene (also known industrially as perchloroethylene, PERC or PCE) and trichloroethene (TCE) by zero-valent iron has been observed to produce cis-1, 2-dichloromethane (DCE) and vinylchloride (VC), both being of considerable toxicological concern.

Decrease in iron reactivity over time, probably due to the formation of surface passive layers or the precipitation of metal hydroxides (e.g. Fe (OH)2, Fe (OH)3) and metal carbonates (e.g., FeCO3) on the surface of iron. Some other metals, especially zinc and tin, can transform HOCs quicker than iron. Palladium, with its superior catalytic ability produced spectacular results as well. For example, recent studies found out that palladized iron can completely dechlorinate many chlorinated aliphatic compounds to hydrocarbons. In some researches, synthesized nanoscale iron and palladized iron particles are used for degradation of chlorinated compounds.

Another metal acting as a catalyst is nickel, Ni(II). This metallic catalyst could prevent formation of toxic by-products by dehalogenation of chlorinated compounds via hydrogen reduction rather than electron transfer. Many researchers have focused on the synthesis of Ni/Fe nanoparticles for the reduction of chlorinated compounds.

One of the major problems associated with ground- and surface-waters is nitrate contamination. The pH value is a means to control the reduction of nitrate by iron and, in effect, the formation of a passive oxide layer.

The deposition of small amounts of a second metal, such as Pd, Pt, Ag, Ni, and Cu, on iron has been shown to accelerate the reaction rate. Whereas iron is deposited with the second metal, a relative potential difference drives the electron from iron to that metal.

Liou et al. used uncatalyzed and catalyzed nanoscale FeO systems for the denitrification of unbuffered 40 mg/L nitrate solutions at initial neutral pH. Compared to microscale FeO (<100 mesh), the efficiency and rate of nitrate removal using uncatalyzed and catalyzed nano-FeO were highly promoted. The maximum elevated rate was obtained using copper-catalyzed nano-FeO (nano-Cu/Fe).

Another synthetic bimetallic nanoparticle is Pd/Au, which reduced the chlorinated compounds from water and ground water. Nutt et al. synthesized Pd supported on gold nanoparticles (Au NPs). They found that these catalysts were considerably more active than Pd NPs. Joo and Zhao [68] prepared Fe-Pd bimetallic with 0.2% w/w of sodium carboxy methyl cellulose (CMC) as stabilizer and used them for the degradation of lindane and atrazine, the chlorinated herbicides.

Nanoparticulate Photocatalysts and Catalysts

Catalysis involves the modification of a chemical reaction rate, mostly speeding up or accelerating the reaction rate by a substance called catalyst that is not consumed throughout the reaction. Usually, the catalyst participates in the reaction by interacting with one or more of the reactants and at the end of process; it is regenerated without any changes.

There are two main kinds of catalysts, homogeneous and heterogeneous. The homogeneous type is dispersed in the same phase as the reactants. The dispersal is ordinarily in a gas or a liquid solution. Heterogeneous catalyst is in a different phase from the reactants and is separated by a phase boundary. Heterogeneous catalytic reactions typically take place on the surface of a solid support, e.g. silica or alumina. These solid materials have very high surface areas that usually arise from their impregnation with acids or coating with catalytically active material e.g. platinum-coated surfaces.

Catalysts usually have two principal roles in nanotechnology areas:

(1) In macro quantities, they can be involved in some processes for the preparation of a variety of other nanostructures like quantum dots, nanotubes, etc.

(2) Some nanostructures themselves can serve as catalysts for certain chemical reactions.

The chemical activity of a conventional heterogeneous catalyst is proportional to its overall specific surface area per unit volume, which is customarily reported in the unit of square metres per grams, with typical values for commercial catalysts in the range of 100 to 400 m^2 /g. There are different procedures to enhance the surface area of the catalyst, which result in voids or empty spaces within the material. It is quite common for these materials to have pores with diameters in the nanometre range.

The pore surface areas are usually determined by the Brunauer-Emmett-Teller (BET) method.

The active component of a heterogeneous catalyst can be a transition ion. Example of some metallic oxides that serve as catalysts, either by themselves or by distribution on a supporting material, are NiO, Cr2O3, Fe2O3, Fe3O4, Co3O4. For some reactions, the catalytic activity arises from the presence of acid sites on the surface. These sites correspond to either Bronsted acids, which are proton donors, or Lewis acids, which are electron pair acceptors.

Pajonk prepared nanoparticles from the sol-gel chemistry combined with the supercritical drying method (aerogels) to enhance the catalyst properties such as textural and thermal qualities. Rare earth metal oxides modification of automotive catalysts (e.g. CeO2, ZrO2) for exhaust gas treatment has resulted in the structural stability, catalytic functions and resistance to sintering at high temperatures [70]. Owing to the low redox potential of non-stochiometrics CeO2, oxygen could release with conversion between 3+and 4+ oxidation states of the Ce ions. This is shown to be essential for effective catalytic functions under the dynamic air-to-fuel ratio cycling.

Nanocrystal surfaces usually are coated with capping ligands. These ligands manage the surface character (both the chemistry and physical structures) which expressed greatly the photocatalytic properties, such as fluorescence lifetimes, quantum yields, surface charge and particle solubility.

The metal complex porphyrins and other biological porphyrin-type molecules like vitamin B12, which are referred to as metalloporphyrinogens, have several characteristics that make them very applicable for the treatment of persistent organic pollutants. They act as redox catalysts for many reactions and known to be active over a long range of redox potentials, electrochemically active with almost any metal, function well in aqueous solutions in the groundwater environment, and highly stable, which enables reactions under severe conditions that prevent other treatment methods (e.g., bioremediation).

Metalloporphyrinogens are molecules with nanometre size and known to catalyze the decomposition of COC (chloro-organic compounds) by reduction reactions. Dror et al. applied these catalysts immobilized in sol-gel matrix, for the reduction of COC. They performed experiments under conditions suitable for ground water systems with titanium citrate and zero-valent iron as electron donors. All the chloro-organic compounds used in these experiments were reduced in the presence of several sol-gel-metalloporphyrinogen hybrids (heterogeneous catalysts).

Wang et al. investigated the effect of size, fabrication method, and morphology of ZnO nanoparticles as photocatalysts on the decomposition of methyl orange. They have used ZnO nanoparticles with different diameters of 10, 50, 200, and 1000 nm. ZnO particles were prepared by two methods of chemical deposition and thermal evaporation. It was found that the preparation method was the most important step and ZnO-nanoparticle, with 50 nm diameter synthesized via thermal evaporation method, provided the highest photocatalyst activity.

Huang et al. used ZnWO4 nanoparticle as photocatalyst for the degradation of rhodamine B in water and decomposition of formaldehyde in gas phase. The sample had the highest photocatalytic activity when prepared at 450°C for 1h. The temperature and time of annealing was observed to be effective for photoatalytic activity. In addition, Lin et al. prepared ZnWO4 nanoparticles and nanorods and used them as photocatalysts for the photodegradation of Rhodamine B and gaseous formaldehyde.

Nanoclays

Clays are layered minerals with space in between the layers where they can adsorb positive and negative ions and water molecules. Clays undergo exchange interactions of adsorbed ions with the outside too.

Although clays are very useful for many applications, they have one main disadvantage i.e. lack of permanent porosity. To overcome this problem, researchers have been looking for a way to prop and support the clay layers with molecular pillars. Most of the clays can swell and thus increase the space in between their layers to accommodate the adsorbed water and ionic species. These clays were employed in the pillaring process.

As stated previously, ultra-fine TiO2 powders have large specific surface areas, but due to their easy agglomeration, an adverse effect on their catalytic performance has been observed. Ding et al. experienced that the recovery of pure TiO2 powders from water was very hard when they used them in aqueous systems. They dispersed TiO2 particles in layered clays and it appeared to provide a feasible solution to such problems. The composite structures, known as pillared clay, could stabilize TiO2 particles and give access of different molecules to the surface of TiO2 crystals. In addition, the interlayer surface of pillared clays is generally hydrophobic, and this is an advantage in adsorption and enriching diluted hydrophobic organic compound in water.

Ooka et al. prepared four kinds of TiO2 pillared clays from different raw clays such as montmorillonite, saponite, fluorine hectorite and fluorine mica. They have tested the surface hydrophobicities and performances of clays in adsorption- photocatalytic decomposition of phthalate esters. It was found that surface hydrophobicity of pillared clays (especially TiO2) largely varied with the host clay. Since the TiO2 particles in the pillared clays are too small to form a crystal phase, they presented a poor photocatalytic activity. To overcome this problem, nanocomposite of titanium dioxide (TiO2) and silicate nanoparticles were made by reaction between titanium hydrate sol of strong acidity and smectite clays in the presence of polyethylene oxide (PEO) surfactants. It resulted in forming larger precursors of TiO2 nanoparticles and condensing them on the fragmentized pieces of the silicate. Introducing PEO surfactants into the synthesis process significantly enhanced the porosity and surface area of the composite solid.

In other works, nanocomposite of iron oxide and silicate was also synthesized for degradation of azo-dye orange (II). To improve the sorption capacity, clays were modified in different ways, such as treatment by inorganic and organic compounds, acids and bases.

Organoclays have recently attracted lots of attention in a number of applications, such as dithiocarbamate-anchored polymer/ organosmectite for the removal of heavy metal ions from aqueous media.

A new class of nano-sized large porous titanium silicate (ETAS-10) and aluminum-substituted ETAS-10 with different Al2O3/TiO2 ratios were successfully synthesized and applied to the removal of heavy metals, in particular Pb2+ and Cd2+. Since tetra-valent Ti is coordinated by octahedral structure, it creates two negative charges that must be normally balanced by two mono-valent cations. This leads to a great interest in ion exchange or adsorption property of this material.

Wang and Wang prepared a series of biopolymer chitosan / montmorillonite (CTS/MMT) nanocomposites and used them as sorbents for the adsorption of Congo Red. They investigated the effect of pH and temperature and found that the sorption capacity was increased with increasing the CTS to MMT ratio.

Dendronized Polymer

Dendronized polymers (or dendronised polymers) are linear polymers to every repeat unit of which dendrons are attached. Dendrons are regularly branched, tree-like fragments and for larger ones the polymer backbone is wrapped to give sausage-like, cylindrical molecular objects. Figure depicted earlier, shows a cartoon representation with the backbone in red and the dendrons like cake slices in green. It also provides a concrete chemical structure showing a polymethylmethacrylate (PMMA) backbone, the methyl group of which is replaced by a dendron of the third generation (three consecutive branching points).

Structure and Applications

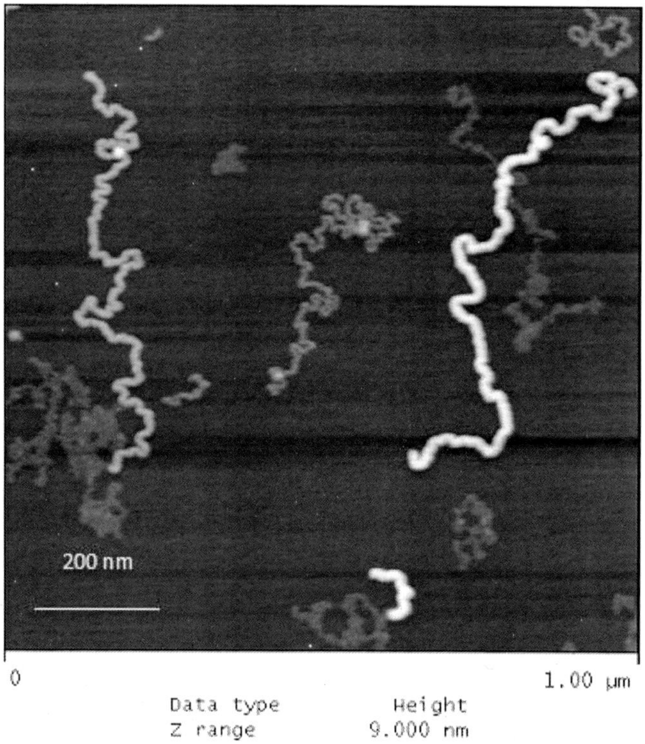

Figure: *Atomic force microscopy height image of co-prepared dendronized polymers of generation one through four (PG1-PG4) reflecting the different thicknesses and apparent persistence lengths for each generation*

Dendronized polymers can contain several thousands of dendrons in one macromolecule and have a stretched out, anisotropic structure. In this regard they differ from the more or less spherically shaped dendrimers, where a few dendrons are attached to a small, dot-like core resulting in an isotropic structure. Depending on dendron generation, the polymers differ in thickness as the atomic force microscopy image shows. Neutral and charged dendronized polymers are highly soluble in organic solvents and in water, respectively. This is due to their low tendency to entangle. Dendronized polymers have been synthesized with, e.g., polymethylmethacrylate, polystyrene, polyacetylene, polyphenylene, polythiophene, polyfluorene, poly(phenylene vinylene), poly(phenylene acetylene), polysiloxane, polyoxanorbornene, poly(ethylene imine)(PEI) backbones. Molar masses up to 200 Mio g/mol have been obtained. Dendronized polymers have been investigated for/as bulk structure control, responsivity to external stimuli, single molecule chemistry, templates for nanoparticle formation, catalysis, electro-optical devices, and bio-related applications. Particularly attractive is the use of water-soluble dendronized polymers for the immobilization of enzymes on solid surfaces (inside glass tubes or microfluidic devices) and for the preparation of dendronized polymer-enzyme conjugates.

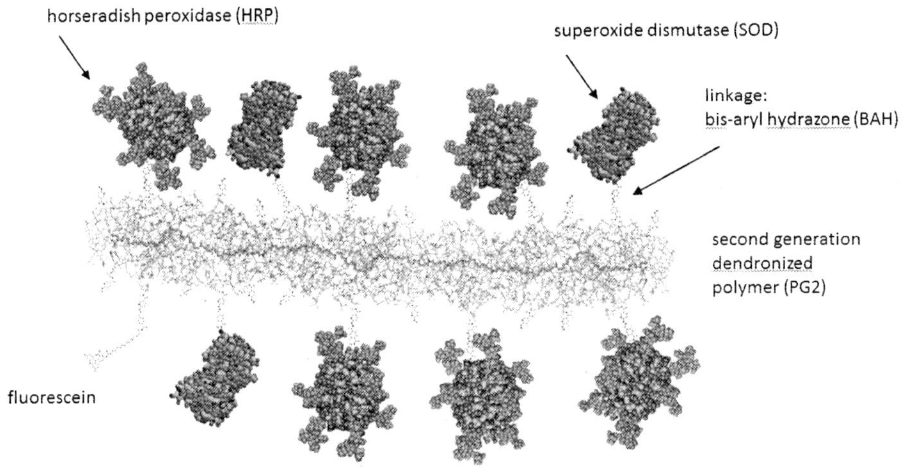

Figure: *Schematic representation of a molecular hybrid structure (conjugate) between a dendronized polymer and the two different enzymes HRP (horseradish peroxidase) and SOD (Cu,Zn-superoxide dismutase). PDB (SOD): 1SXA; PDB (HRP): 1ATJ; HRP sugar modification from Gray & Montgomery,* Carbohydrate Research, *2006,* 341, *198-209. Denpol structure: Bertran et al.* RSC Adv., *2013,* 3, *126-140.*

Synthesis

The two main approaches into this class of polymers are the macromonomer route and the attach-to route. In the former, a monomer which already carries the dendron of final size is polymerized. In the latter the dendrons are constructed generation by generation directly on an already existing polymer. The macromonomer route results in shorter chains for higher generations and the attach-to route is prone to lead to structure imperfections as an enormous number of chemical reactions have to be performed for each macromolecule.

History

The name "Dendronized Polymer" which meanwhile is internationally accepted was coined by Schlüter in 1998. The first report on such a macromolecule which at that time was called "Rod-shaped Dendrimer" goes back to a patent by Tomalia in 1987 and was followed by Percec's first mentioning in the open literature of a polymer with "tapered side chains" in 1992. In 1994 the potential of these polymers as cylindrical nanoobjects was recognised. Many groups worldwide contributed to this field.

Metallodendrimer

A Metallodendrimer is a type of dendrimer with incorporated metal atoms. The development of this type of material is actively pursued in academia.

Structure

The metal can be situated in the repeat unit, the core or at the extremities as end-group. Elements often encountered are palladium and platinum. These metals can form octahedral six-coordinate M(IV) linking units from organic dihalides and the corresponding 4-coordinate M(II) monomers. Ferrocene-containing dendrimers and dendrimers with cobaltocene and arylchromiumtricarbonyl units have been reported in end-functional dendrimers.

Metallodendrimers can form as metal complexes with dendritic counter ions for example by hydrolysis of ester terminated PAMAM dendrimers with sodium hydroxide.

Applications

Metallodendrimers are investigated as equivalents to nanoparticles. Applications can be expected in the fields of catalysis, as chemical sensors in molecular recognition - for example of bromine and chloride anions - or as materials capable of binding metals. Metallodendrimers can also mimic certain biomolecules for example haemoprotein in dendrimer with a porphyrin core. Further uses are reported as electrocatalyst.

Examples of metallodendrimer heterogeneous catalysis are a nickel-containing dendrimer active in the Kharasch addition, palladium-containing dendrimers active in ethylene polymerization and in the Heck reaction.

Ferrocene-Containing Dendrimers

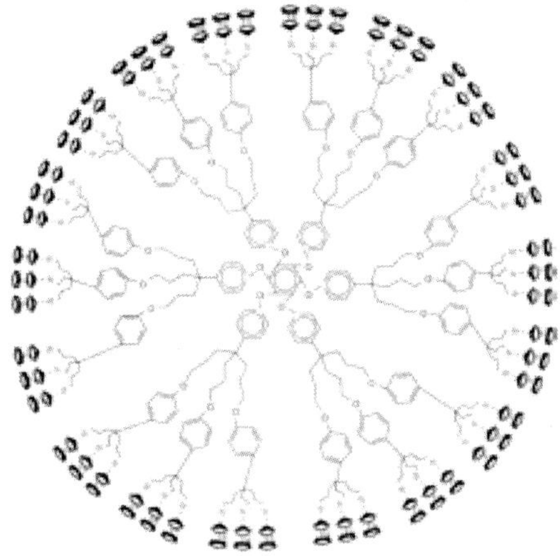

Figure: *A 54-ferrocene dendrimer*

Ferrocene-containing dendrimers are of interest for their ability to do multielectron-transfers and they have been synthesized by convergent and divergent approaches. There are three basic categories that ferrocene-containing dendrimers can be placed in: (a) dendrimers with ferrocene cores, (b) dendrimers with peripheral ferrocene groups, and (c) dendrimers with ferrocene at the core and the peripheral.

Synthesis

Ferrocene-containing dendrimers can be synthesized by both convergent and divergent methods. Some of the first dendrimers of this type, were made by attaching ferrocene units to small silicon containing dendrimers.

Dendrimers with peripheral ferrocene groups are usually synthesized by attaching ferrocene to the core by either olefin metathesis or by hydrosilylation reactions. As an example, compound 1 was synthesized by a divergent approach in which tetraallylsilane undergoes Pt-catalyzed hydrosilylation to form the core. This core was then reacted with ferrocenyllithium to form 1. Convergent approaches can also be used to make dendrimers with peripheral ferrocene. As an example, figure depicted earlier shows a 54-ferrocene dendrimer which was synthesized by a fast convergent approach.

Dendrimers with ferrocene cores can be synthesized by first making ferrocene derivatives with functional groups. An example of this would be compounds 2 and 3. These compounds can undergo hydrosilylation reactions to form dendrimers. As an example, decaallylferrocene (3) can undergo hydrosilylation with dimethylferrocenylsilane to form a decaferrocenyl functionalized ferrocene as is shown in figure above.

Properties and Applications

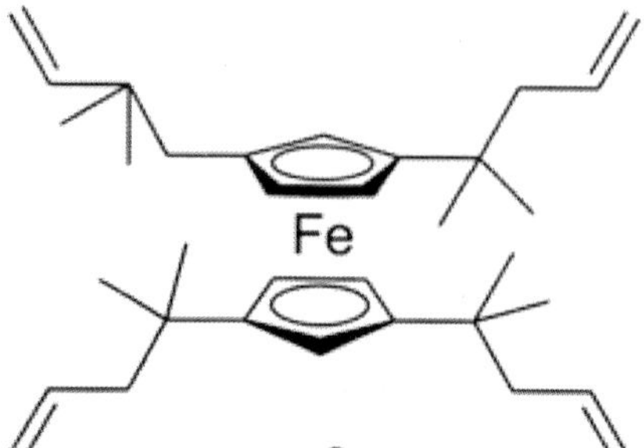

2

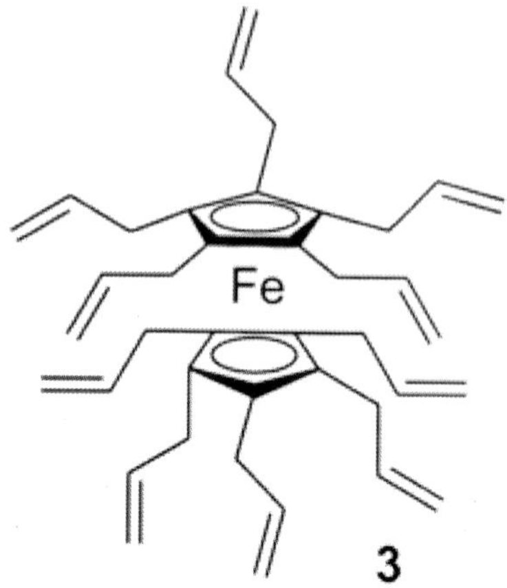

3

Figure: *An asymmetric dendrimer with a ferrocene core*

There are many potential applications for ferrocene containing dendrimers. The CV of the 54-ferrocene containing dendrimer in figure gave a single reversible wave. The ferrocenyl moieties on these dendrimers are essentially noninteracting redox centres. Because of this, they could potentially be used as molecular batteries or sensors. There have been some efforts to synthesize dendrimers that behave similar to redox proteins. Asymmetric dendrimers similar to the one in figure have been synthesized and have undergone inclusion complexation by beta-cyclodextrin.

Nanosponges

A sponge is a tool or cleaning aid consisting of porous material. Sponges are usually used for cleaning impervious surfaces. They are especially good at absorbing water and water-based solutions.

Sponges are commonly made from cellulose wood fibres or foamed plastic polymers. Some natural sponges are still sold, but most are now used either as body or facial sponges (bath sponges) or as tools for sponge painting.

The three other categories of available synthetic sponges are: low-density polyether (known as the rainbow packs of non-absorbent sponges), PVA a highly absorbent material and polyester.

Polyester sponges are subdivided into a variety of types, some of which are reticulated (artificially broken-in) for ease of use. One type, double-blown polyester, has high water-retention ability approaching or equaling that of PVA sponges, but with visible pores and more diverse uses.

Because it is made primarily made of wood fibre, a cellulose sponge can be a medium for the growth of harmful bacteria or fungi, especially when it is allowed to remain wet between uses.

Nano-sponge filter is a high performance filter with wide range of use for aquarium fishes such as guppy, neon, shrimp and all kinds of fish fries. Nano-sponge with fine porous poles maximised the colonization of beneficial bacteria which will help to keep water clear.

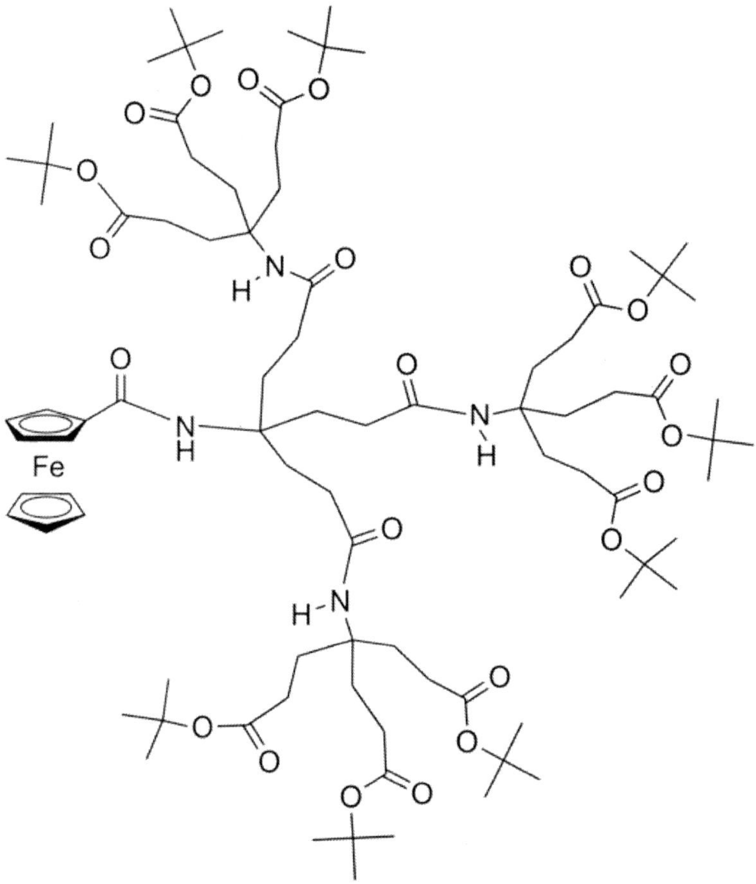

Figure: *A new metal organic framework can effectively filter toxins from water.*

New research has demonstrated the potential of a new kind of nanomaterial to filter out environmental toxins in water. Metal Organic Framework (MOF) are clusters of metal atoms connected by organic molecules and known for their exceptional abilities to store or separate gases such as carbon dioxide.

Researchers at Rice University and Penn State University have discovered that adding a dash of boron to carbon while creating nanotubes turns them into solid, spongy, reusable blocks that have an astounding ability to absorb oil spilled in water. Bacteria are becoming more resistant to the best weapon we have against them: antibiotics. But what if we could live together in peace? Instead of killing the germs, there may be a way to render them harmless to the body using nanosponges to soak up the toxins that make us sick.

The toxin tested was alpha-haemolysin from staphylococcus bacteria. It damages cells because it is what is called "pore forming." The alpha-haemolysin molecules are attracted to the proteins in cell walls, and when they attach, they weaken the molecular bonds — essentially putting holes in the cells. The cell dies, and enough of the toxin is fatal.

Since the inside of the nanosponge is a polymer, and solid, the toxic molecule just sticks there and doesn't damage living cells. At that point, the nanosponge can carry the toxin to the liver, where it gets eliminated from the body. (The alpha-haemolysin doesn't attack the liver cells because the side of the molecule that would damage them is attached to the sponge already). One nanosponge can soak up a lot of alpha-haemolysin molecules.

Currently, however most antidotes to toxins aren't quite so wide-ranging: snake antivenins, for example, are often specific to certain species or families. And snake antivenin won't do any good against the poisons produced by bacterial infections.

With nanosponges designed as red blood cells, even though the venom from some snake species and the bacterial alpha-haemolysin will be different molecules, with vastly different shapes — they work the same way, via their attraction to cell membranes.

Micelles (Self-Assembled Surfactants)

A micelle (/majÈs[l/) or micella (/majÈs[lY/) (plural micelles or micellae, respectively) is an aggregate of surfactant molecules dispersed in a liquid colloid. A typical micelle in aqueous solution forms an aggregate with the hydrophilic "head" regions in contact with surrounding solvent, sequestering the hydrophobic single-tail regions in the micelle centre. This phase is caused by the packing behaviour of single-tail lipids in a bilayer. The difficulty filling all the volume of the interior of a bilayer, while accommodating the area per head group forced on the molecule by the hydration of the lipid head group, leads to the formation of the micelle. This type of micelle is known as a normal-phase micelle (oil-in-water micelle). Inverse micelles have the head groups at the centre with the tails extending out (water-in-oil micelle). Micelles are approximately spherical in shape. Other phases, including shapes such as ellipsoids, cylinders, and bilayers, are also possible. The shape and size of a micelle are a function of the molecular geometry of its surfactant molecules and solution conditions such as surfactant concentration, temperature, pH, and ionic strength. The process of forming micelles is known as micellisation and forms part of the phase behaviour of many lipids according to their polymorphism.

Micelles are self-assembled surfactant materials in a bulk solution.

Surfactants or "Surface active agents" are usually organic compounds that are amphipathic, meaning they contain both hydrophobic groups (tails) and hydrophilic groups (heads). Therefore, they are typically soluble in both organic solvents and water.

There are hundreds of compounds that can be used as surfactants and are usually classified by their ionic behaviour in solutions; anionic, cationic, non-ionic or amphoteric (zwiterionic). Each surfactant class has its own specific properties. A surfactant can be classified by the presence of formally charged groups in its head. There are no charge groups in a head of nonionic surfactant. The head of an ionic surfactant carries a net charge. If the charge is negative, the surfactant is more specifically called anionic and if the charge is positive, it is called cationic. If a surfactant contains a head with two oppositely charged groups, it is termed zwitterionic.

The concentration at which surfactants begin to self-assemble and form micelles is known as critical micelle concentration or CMC. When micelles appear in the water, their tails form a core that is like an oil droplet, and their (ionic) heads form an outer shell that maintains favourable contact with water. The self-assembled surfactant is referred to as "reverse micelle" when surfactants assemble in the oil. In this case the heads are in the core and the tails have favourable contact with oil.

Surfactant-enhanced remediation techniques have shown significant potential in their application for the removal of polycyclic aromatic hydrocarbon (PAHs) pollutants in the soil. Increasing surfactant concentration in the solution has shown higher effectiveness in the extraction of NAPLs (non-aqueous phase liquids) and PAHs. At high concentrations, surfactant solutions improve the formation of pollutant emulsions that are hard to extract from the sample.

On the other hand, surfactant solutions with low concentrations are not very effective in solubilizing the pollutants. As a result, recent research has been directed towards the design of a surfactant that minimizes their losses and the development of surfactant recovery and recycling techniques. To overcome these problems, Kim et al. tested amphiphilic polyurethane (APU) nano-network polymer particles. They examined the APU efficiency to remove a model hydrophobic pollutant (phenantrene) from a contaminated sandy aquifer material. One of the advantages of the APU particle emulsion is the wide range of concentration that can be used in soil remediation. APU nano-network suspensions extracted up to 98% of the phenanthrene adsorbed on the aquifer material with extremely low loss of particles.

Self-Assemblies

Self-assembly is defined as a reversible process in which pre-existing parts, or disordered components of a pre-existing system, form structures of patterns. In other words, a self-assembly process is the spontaneous organisation of small molecules into larger well-defined stable, ordered molecular complexes or aggregates, and spontaneous adsorption of atoms or molecules onto a substrate in a systematic ordered manner.

The most well-studied subfield of self-assembly is molecular self-assembly, but in recent years it has been demonstrated that self-assembly is possible with micro and millimetre scale structures lying in the interface between two liquids like micelles as discussed before.

Molecular self-assembly is gathering of molecules without guidance or management from an outside source. There are two types of self-assembly i.e. intramolecular and intermolecular, although the term self-assembly itself usually refers to intermolecular one. Intramolecular self-assembling molecules are complex polymers with the ability to assemble from the random coil conformation into a well-defined stable structure (secondary and tertiary structure). An example of this type of self-assembly is protein folding. Intermolecular self-assembly is the ability of molecules to form supramolecular assemblies (quaternary structure). A simple example is the formation of a micelle by surfactant molecules in solution as discussed before.

Attaching a monolayer of molecules to mesoporous ceramic supports gives materials known as Self-Assembled Monolayers on Mesoporous

Supports (SAMMS). The highly ordered nanostructure of SAMMS is the result of three molecular-self-assembly stages. The first stage is the aggregation of the surfactant molecules to make the micelle template. The second generation is the aggregation of the silicate-coated micelles into the mesostructured body, and the third is the self assembly of the silane molecules into an ordered monolayer structure across the pore interface.

The resulted functionalized hexagonal structure is a base to build an environmental sorbent material. The rigid, open pore structure of the supports makes all of the interfacial binding sites accessible to solution species, and results in fast sorption kinetics. All the above advantages of SAMMS make it ready for various chemical bindings. SAMMS of silica-based materials are highly efficient sorbents for target species, such as heavy metals, tetrahedral oxometalate anions, and radionuclides. The self-assembly technique is also used for preparation of catalysts in the form of thin films on a support material. Some of these thin film catalysts have great uniformities and high photocatalytic activities, but their thicknesses can hardly be controlled. Thus, preparing them in large areas is rather difficult.

A suitable technique of preparing ultra-thin films with precise thickness adjustment is the layer-by-layer self-assembly procedure.

The principle of multilayer assembly is quite simple; colloidal particles will self-assemble on the surface of a suitable solid substrate because of its surface forces. In most cases, electrostatic interaction provides the stability for the films.

Szabo et al. provided a new process for synthesizing $Zn(OH)_2$ and ZnO nanoparticles. For this purpose, multilayer films of $Zn(OH)_2$ and ZnO nanoparticles were prepared by the layer-by-layer self-assembly technique on glass surface. Photocatalytic measurements were made with model organic materials β-naphtol and industrial kerosene in a loop-type batch reactor.

Nanomembrane and Nanosieve

A membrane is a semi-permeable and selective barrier between two phases (retentive and permeate) through which only selected chemical species may diffuse. Membrane filtration is frequently employed for the separation of dissolved solutes in a fluid or the separation of a gas mixture.

Historically, membrane technology has had wide application in wastewater treatment and desalination via reverse osmosis. In this method, a pressure difference across a membrane is employed to overcome the osmotic pressure gradient. The smaller water molecules are literally pushed through the membrane while the large solute species are retained behind.

Among different classes of membranes, reverse osmosis (RO) filtration is a well known process in the desalination of seawater and ultrafiltration (UF) is a well established process in the fractionation of Natural Organic Matter (NOM). Nanofiltration (NF) is a process with membrane permeability between RO and UF. Another membrane design is emulsion liquid membrane (ELM). An ELM is formed by first encapsulating an aqueous "receiving" or strip phase within a hydrophobic membrane liquid. This emulsion is then further dispersed within the continuous aqueous feed phase. This technology was used for the extraction of phenols, removal of heavy metal cations such as zinc, cadmium, chromium, copper, lead, palladium and mercury from wastewater and also removal of alkali metal cations such as Na+, K+, Li+ and Cs+, radioactive fission products, such as Cs-137, Sr-90, Ce-139 and Eu-152 and anions, such as chlorides, sulfate, phosphate and chromate.

The efficiency of nitrate removal by three commercial nanofiltration membranes, NF90, NF270 (Dow-FilmTec) and ESNA1-LF (Hydranautics) was studied by Santafe-Moros.

Natural organic matters (NOM) such as humic acid and fulvic acid are widely distributed in soil, natural water and sediments. These materials include mixture of the degradation products of plant and animal residues. Membrane 2TiO-modified photocatalytic oxidation (PCO) has been focused on the NOM removal and/or destruction processes.

Fu et al. designed a submerged membrane photocatalysis reactor (SMPR) for degradation of fulvic acid via novel nano-structured 2TiO/ silica gel photocatalyst. The possibility of using novel TiO2 for the prevention of microfilteration membrane fouling for water purification was tested as well.

Ultrafiltration has been applied in most membrane separation processes. The hydrophilicity of the membrane and its porous structure play important roles in these processes. An appropriate porous membrane must have high permeability, good hydrophilicity and excellent chemical resistance to the feed streams. In order to obtain high permeability, membranes should have high surface porosity, and good pore structure.

Polyvinylidene fluoride (PVDF) is a material that can form such asymmetric membranes, since it is thermally stable and resistant to corrosion by most chemicals and organic compounds. PVDF-based membranes exhibit outstanding anti-oxidation activities, strong thermal and hydrolytic stabilities and good mechanical properties. Yan et al. studied the modification of polyvinylidene fluoride (PVDF) ultrafiltration membrane by dispersing nano-sized alumina (32OAl) particles uniformly in a PVDF solution (19% polymer weight).

Using selective polymeric membranes for gas separation is also a promising process. Efficient gas separation membranes are required to have both high permeability and selectivity. Compared with flat membranes, hollow fibre membranes are more favoured due to a larger membrane area per volume, good flexibility and easy handling in the module fabrication. Mixed matrix composite membranes that are fabricated by encapsulating the molecular sieves into the polymer matrix have also been recognised. The hollow fibres are developed with a thin zeolite beta-polysulfone mixed matrix selective layer and improved selectivity for He/N2 and O2/N2 separation.

Ultrafiltration

Ultrafiltration (UF) is a variety of membrane filtration in which forces like pressure or concentration gradients leads to a separation through a semipermeable membrane. Suspended solids and solutes of high molecular weight are retained in the so-called retentate, while water and low molecular weight solutes pass through the membrane in the permeate. This separation process is used in industry and research for purifying and concentrating macromolecular (10^3 - 10^6 Da) solutions, especially protein solutions. Ultrafiltration is not fundamentally different from microfiltration, nanofiltration or membrane gas separation, except in terms of the size of the molecules it retains - it is defined by the Molecular Weight Cut Off (MWCO) of the membrane used. Ultrafiltration is applied in cross-flow or dead-end mode.

Industries such as chemical and pharmaceutical manufacturing, food and beverage processing, and waste water treatment, employ ultrafiltration in order to recycle flow or add value to later products. But also blood dialysis belongs to ultrafiltration.

Drinking Water

Figure: *Drinking water treatment 300 m^3/h using ultrafiltration in Grundmühle waterworks (Germany)*

UF can be used for the removal of particulates and macromolecules from raw water to produce potable water. They have been used to either replace existing secondary (coagulation, flocculation, sedimentation) and tertiary filtration (sand filtration and chlorination) systems employed in water treatment plants or as standalone systems in isolated regions with growing populations. When treating water with high suspended solids, UF is often integrated into the process, utilising primary (screening, flotation, filtration) and some secondary treatments as pre-treatment stages. UF processes are currently preferred over traditional treatment methods for the following reasons:

- No chemicals required (aside from cleaning)
- Constant product quality regardless of feed quality
- Compact plant size
- Capable of exceeding regulatory standards of water quality, achieving 90-100% pathogen removal

UF processes are currently limited by the high cost incurred due to membrane fouling and replacement. Additional pretreatment of feed water is required to prevent excessive damage to the membrane units.

In many cases UF is used for pre filtration in reverse osmosis plants to protect the RO.

Protein Concentration

UF is used extensively in the dairy industry; particularly in the processing of cheese whey to obtain whey protein concentrate (WPC) and lactose-rich permeate. In a single stage, a UF process is able to concentrate the whey 10-30 times the feed.

The original alternative to membrane filtration of whey was using steam heating followed by drum drying or spray drying. The product of these methods had limited applications due to its granulated texture and insolubility. Existing methods also had inconsistent product composition, high capital and operating costs and due to the excessive heat used in drying would often denature some of the proteins.

Compared to traditional methods, UF processes used for this application:

- Are more energy efficient
- Have consistent product quality, 35-80% protein product depending on operating conditions
- Do not denature proteins as they use moderate operating conditions

The potential for fouling is widely discussed, being identified as a significant contributor to decline in productivity. Cheese whey contains high concentrations of calcium phosphate which can potentially lead to scale deposits on the membrane surface. As a result substantial pretreatment must be implemented to balance pH and temperature of the feed to maintain solubility of calcium salts.

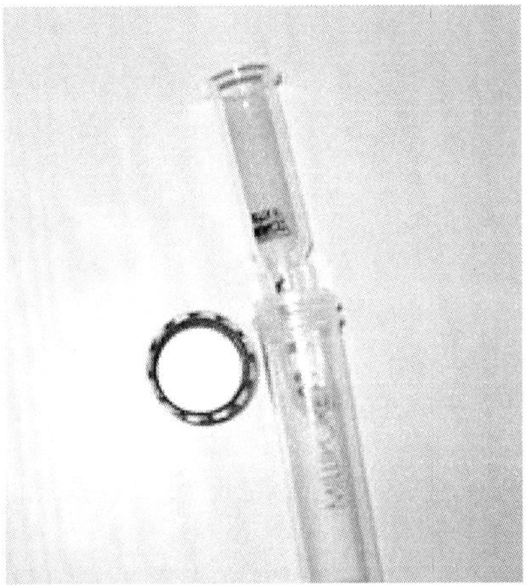

Figure: *A selectively permeable membrane can be mounted in a centrifuge tube. The buffer is forced through the membrane by centrifugation, leaving the protein in the upper chamber.*

Other Applications

- Filtration of effluent from paper pulp mill
- Cheese manufacture
- Removal of pathogens from milk
- Process and waste water treatment
- Enzyme recovery
- Fruit juice concentration and clarification
- Dialysis and other blood treatments
- Desalting and solvent-exchange of proteins (via diafiltration)
- Laboratory grade manufacturing

Principles

The basic operating principle of ultrafiltration uses a pressure induced separation of solutes from a solvent through a semi permeable membrane. The relationship between the applied pressure on the solution to be separated and the flux through the membrane is most commonly described by the Darcy equation:

$$J = \frac{TMP}{\mu R_t}$$

where J is the flux (flow rate per membrane area), TMP is the transmembrane pressure (pressure difference between feed and permeate stream), μ is solvent viscosity, R_t is the total resistance (sum of membrane and fouling resistance).

Membrane Fouling

Concentration Polarization

When filtration occurs the local concentration of rejected material at the membrane surface increases and can become saturated. In UF, increased ion concentration can develop an osmotic pressure on the feed side of the membrane. This reduces the effective TMP of the system, therefore reducing permeation rate. It must be noted that concentration polarization differs to fouling as it has no lasting effects on the membrane itself and can be reversed by relieving the TMP. It does however have a significant effect on many types of fouling.

The Types of Fouling

Particulate Deposition

The following models describe the mechanisms of particulate deposition on the membrane surface and in the pores:

- *Standard blocking*: macromolecules are uniformly deposited on pore walls
- *Complete blocking*: membrane pore is completely sealed by a macromolecule
- *Cake filtration*: accumulated particles or macromolecules form a fouling layer on the membrane surface, in UF this is also known as a gel layer
- *Intermediate blocking*: when macromolecules deposit into pores or onto already blocked pores, contributing to cake formation

Scaling

As a result of concentration polarization at the membrane surface, increased ion concentrations may exceed solubility thresholds and precipitate on the membrane surface. These inorganic salt deposits can block pores causing flux decline, membrane degradation and loss of production. The formation of scale is highly dependent on factors affecting both solubility and concentration polarization including pH, temperature, flow velocity and permeation rate.

Biofouling

Microorganisms will adhere to the membrane surface forming a gel layer – known as biofilm. The film increases the resistance to flow, acting as an additional barrier to permeation. In spiral-wound modules, blockages formed by biofilm can lead to uneven flow distribution and thus increase the effects of concentration polarization.

Membrane Arrangements

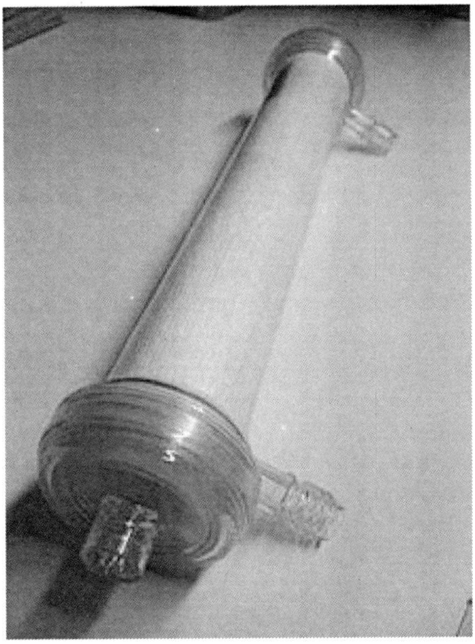

Figure: *Hollow fibre module*

Depending on the shape and material of the membrane, different modules can be used for ultrafiltration process. Commercially available designs in ultrafiltration modules vary according to the required hydrodynamic and economic constraints as well as the mechanical stability of the system under particular operating pressures. The main modules used in industry include:

Tubular Modules

The tubular module design uses polymeric membranes cast on the inside of plastic or porous paper components with diameters typically in the range of 5 – 25 mm with lengths from 0.6 - 6.4 m. Multiple tubes are housed in a PVC or steel shell. The feed of the module is passed through the tubes, accommodating radial transfer of permeate to the shell side. This design allows for easy cleaning however the main drawback is its low permeability, high volume hold-up within the membrane and low packing density.

Hollow Fibre

This design is conceptually similar to the tubular module with a shell and tube arrangement. A single module can consist of 50 to thousands of hollow fibres and therefore are self-supporting unlike the tubular design. The diameter of each fibre ranges from 0.2 – 3 mm with the feed flowing in the tube and the product permeate collected radially on the outside. The advantage of having self-supporting membranes as is the ease at which it can be cleaned due to its ability to be backflushed. Replacement costs however are high, as one faulty fibre will require the whole bundle to be replaced. Considering the tubes are of small diameter, using this design also makes the system prone to blockage.

Spiral-Wound Modules

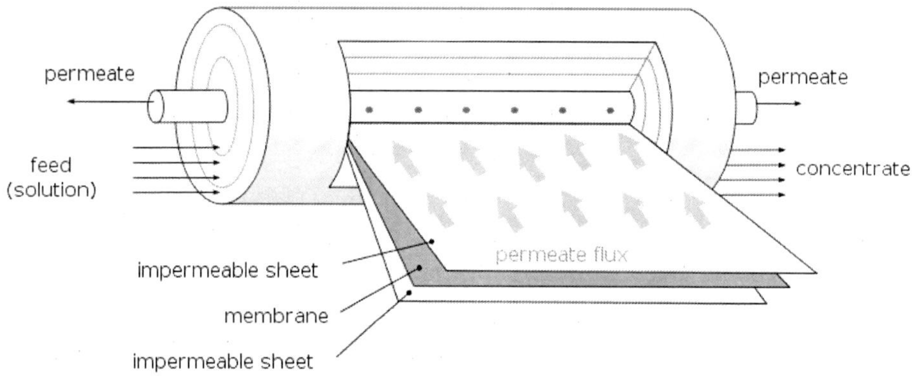

Figure: *Spiral-wound membrane module*

Are composed of a combination of flat membrane sheets separated by a thin meshed spacer material which serves as a porous plastic screen support. These sheets are rolled around a central perforated tube and fitted into a tubular steel pressure vessel casing. The feed solution passes over the membrane surface and the permeate spirals into the central collection tube. Spiral-wound modules are a compact and cheap alternative in ultrafiltration design, offer a high volumetric throughput and can also be easily cleaned. However it is limited by the thin channels where feed solutions with suspended solids can result in partial blockage of the membrane pores.

Plate and Frame

This uses a membrane placed on a flat plate separated by a mesh like material. The feed is passed through the system from which permeate is separated and collected from the edge of the plate. Channel length can range from 10 – 60 cm and channel heights from 0.5 – 1 mm. This module provides low volume hold-up, relatively easy replacement of the membrane and the ability to feed viscous solutions because of the low channel height, unique to this particular design.

Process Characteristics

The process characteristics of a UF system are highly dependent on the type of membrane used and its application.

Manufacturers' specifications of the membrane tend to limit the process to the following typical specifications:

	Hollow Fibre	*Spiral-Wound*	*Ceramic Tubular*
pH	2-13	2-11	3-7
Feed Pressure (psi)	9-15	<30-120	60-100
Backwash Pressure (psi)	9-15	20-40	10-30
Temperature (°C)	5-30	5-45	5-400
Total Dissolved Solids (mg/L)	<1000	<600	<500
Total Suspended Solids (mg/L)	<500	<450	<300
Turbidity (NTU)	<15	<1	<10
Iron (mg/L)	<5	<5	<5
Oils and Greases (mg/L)	<0.1	<0.1	<0.1
Solvents, phenols (mg/L)	<0.1	<0.1	<0.1

Process Design Considerations

When designing a new membrane separation facility or considering its integration into an existing plant, there are many factors which must be considered. For most applications a heuristic approach can be applied to determine many of these characteristics to simplify the design process. Some design areas include:

Pre-Treatment

Treatment of feed prior to the membrane is essential to prevent damage to the membrane and minimize the effects of fouling which greatly reduce the efficiency of the separation.

Types of pre-treatment are often dependent on the type of feed and its quality. For example in wastewater treatment, household waste and other particulates are screened. Other types of pre-treatment common to many UF processes include pH balancing and coagulation.

Appropriate sequencing of each pre-treatment phase is crucial in preventing damage to subsequent stages. Pre-treatment can even be employed simply using dosing points.

Membrane Specifications

Material

Most UF membranes use polymer materials (polysulfone, polypropylene, cellulose acetate) however ceramic membranes are used for high temperature applications.

Pore Size

A general rule for choice of pore size in a UF system is to use a membrane with a pore size one tenth that of the particle size to be separated. This limits the amount of smaller particles entering the pores and adsorbing to the pore surface. Instead they block the entrance to the pores allowing simple adjustments of cross-flow velocity to dislodge them.

Operation Strategy

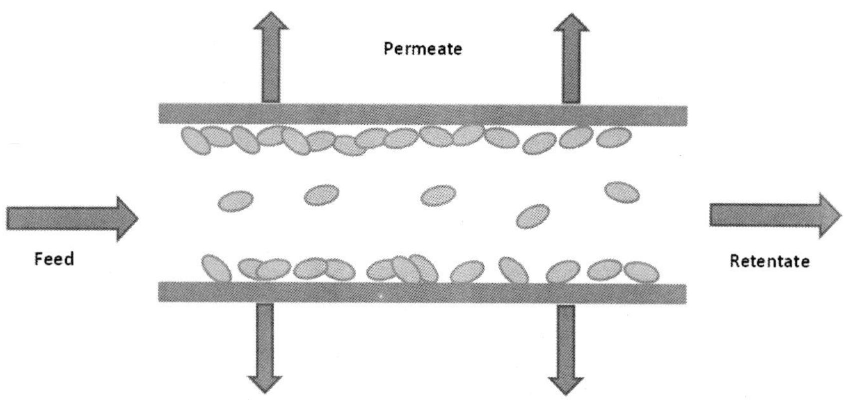

Figure: *Schematic of cross flow operation.*

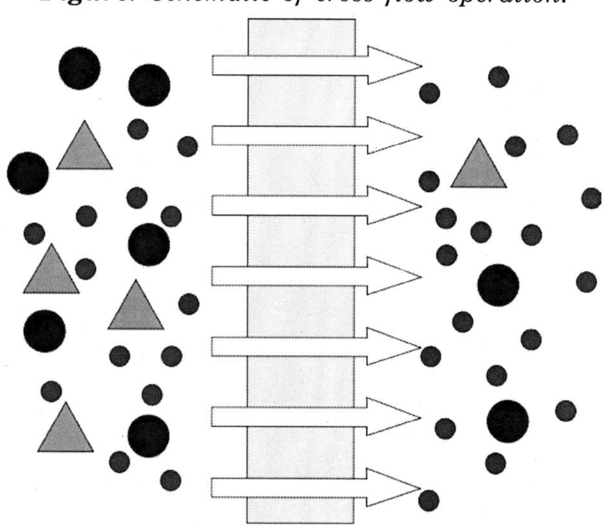

Figure: *Schematic of dead-end operation*

Flow Type

UF systems can either operate with cross-flow or dead-end flow. In dead-end filtration the flow of the feed solution is perpendicular to the membrane surface. On the other hand in cross flow systems the flow passes parallel to the membrane surface.

Dead-end configurations are more suited to batch processes with low suspended solids as solids accumulate at the membrane surface therefore requiring frequent backflushes and cleaning to maintain high flux. Cross-flow configurations are preferred in continuous operations as solids are continuously flushed from the membrane surface resulting in a thinner cake layer and lower resistance to permeation.

Flow Velocity

Flow velocity is especially critical for hard water or liquids containing suspensions in preventing excessive fouling. Higher cross-flow velocities can be used to enhance the sweeping effect across the membrane surface therefore preventing deposition of macromolecules and colloidal material and reducing the effects of concentration polarization. Expensive pumps are however required to achieve these conditions.

Flow Temperature

To avoid excessive damagee to the membrane, it is recommended to operate a plant at the temperature specified by the membrane manufacturer. In some instances however temperatures beyond the recommended region are required to minimise the effects of fouling. Economic analysis of the process is required to find a compromise between the increased cost of membrane replacement and productivity of the separation.

Pressure

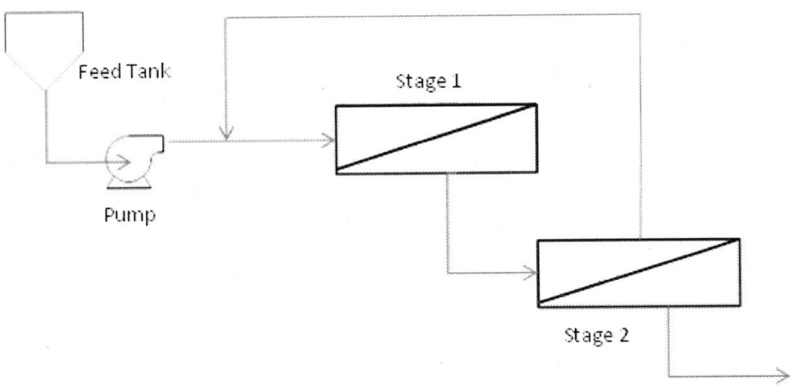

Figure: *Typical two stage membrane process with recycle stream*

Pressure drops over multi-stage separation can result in a drastic decline in flux performance in the latter stages of the process. This can be improved using booster pumps to increase the TMP in the final stages. This will incur a greater capital and energy cost which will be offset by the improved productivity of the process. With a multi-stage operation, retentate streams from each stage are recycled through the previous stage to improve their separation efficiency.

Multi-Stage, Multi-Module

Multiple stages in series can be applied to achieve higher purity permeate streams. Due to the modular nature of membrane processes, multiple modules can be arranged in parallel to treat greater volumes.

Post-Treatment

Post-treatment of the product streams is dependent on the composition of the permeate and retentate and its end-use or government regulation. In cases such as milk separation both streams (milk and whey) can be collected and made into useful products. Additional drying of the retentate will produce whey powder.

In the paper mill industry, the retentate (non-biodegradable organic material) is incinerated to recover energy and permeate (purified water) is discharged into waterways. It is essential for the permeate water to be pH balanced and cooled to avoid thermal pollution of waterways and altering its pH.

Cleaning

Cleaning of the membrane is done regularly to prevent the accumulation of foulants and reverse the degrading effects of fouling on permeability and selectivity.

Regular backwashing is often conducted every 10 min for some processes to remove cake layers formed on the membrane surface. By pressurising the permeate stream and forcing it back through the membrane, accumulated particles can be dislodged, improving the flux of the process. Backwashing is limited in its ability to remove more complex forms of fouling such as biofouling, scaling or adsorption to pore walls.

These types of foulants require chemical cleaning to be removed. The common types of chemicals used for cleaning are:

- Acidic solutions for the control of inorganic scale deposits
- Alkali solutions for removal of organic compounds
- Biocides when bio-fouling is evident

When designing a cleaning protocol it is essential to consider:

Cleaning time – Adequate time must be allowed for chemicals to interact with foulants and permeate into the membrane pores. However if the process is extended beyond its optimum duration it can lead to denaturation of the membrane and deposition of removed foulants. The complete cleaning cycle including rinses between stages may take as long as 2 hours to complete.

Aggressiveness of chemical treatment – With a high degree of fouling it may be necessary to employ aggressive cleaning solutions to remove fouling material. However in some applications this may not be suitable if the membrane material is sensitive, leading to enhanced membrane ageing.

Disposal of cleaning effluent – The release of some chemicals into wastewater systems may be prohibited or regulated therefore this must be considered. For example the use of phosphoric acid may result in high levels of phosphates entering water ways and must be monitored and controlled to prevent eutrophication.

Summary of common types of fouling and their respective chemical treatments

Foulant	***Reagent***	***Time and Temperature***	***Mode of Action***
Fats and oils, proteins, polysaccharides, bacteria	0.5M NaOH with 200 ppm Cl2	30-60 min 25-55°C	Hydrolysis and oxidation
DNA, mineral salts	0.1M – 0.5M acid (acetic, citric, nitric)	30-60 min 25-35°C	Solubilization
Fats, oils, biopolymers, proteins	0.1% SDS, 0.1% Triton X-100	30 min – overnight 25-55°C	Wetting, emulsifying, suspending, dispersing
Cell fragments, fats, oils, proteins	Enzyme detergents	30 min – overnight 30 – 40°C	Catalytic breakdown
DNA	0.5% DNAase	30 min – overnight 20 – 40°C	Enzyme hydrolysis

New Developments

In order to increase the life-cycle of membrane filtration systems, energy efficient membranes are being developed in membrane bioreactor systems. Technology has been introduced which allows the power required to aerate the membrane for cleaning to be reduced whilst still maintaining a high flux level. Mechanical cleaning processes have also been adopted using granulates as an alternative to conventional forms of cleaning; this reduces energy consumption and also reduces the area required for filtration tanks.

Membrane properties have also been enhanced to reduce fouling tendencies by modifying surface properties. This can be noted in the biotechnology industry where membrane surfaces have been altered in order to reduce the amount of protein binding. Ultrafiltration modules have also been improved to allow for more membrane for a given area without increasing its risk of fouling by designing more efficient module internals.

The current pre-treatment of seawater desulphination uses ultrafiltration modules that have been designed to withstand high temperatures and pressures whilst occupying a smaller footprint. Each module vessel is self supported and resistant to corrosion and accommodates easy removal and replacement of the module without the cost of replacing the vessel itself.

6

Mesoporous Oxide Layers Applied to Advanced Energy Conversion Methods

A mesoporous material is a material containing pores with diameters between 2 and 50 nm.

Porous materials are classified into several kinds by their size. According to IUPAC notation, microporous materials have pore diameters of less than 2 nm and macroporous materials have pore diameters of greater than 50 nm; the mesoporous category thus lies in the middle.

Typical mesoporous materials include some kinds of silica and alumina that have similarly-sized fine mesopores. Mesoporous oxides of niobium, tantalum, titanium, zirconium, cerium and tin have also been reported. According to the IUPAC, a mesoporous material can be disordered or ordered in a mesostructure.

A procedure for producing mesoporous materials (silica) was patented around 1970. It went almost unnoticed and was reproduced in 1997. Mesoporous silica nanoparticles (MSNs) were independently synthesized in 1990 by researchers in Japan. They were later produced also at Mobil Corporation laboratories and named Mobil Crystalline Materials, or MCM-41.

Since then, research in this field has steadily grown. Notable examples of prospective applications are catalysis, sorption, gas sensing, ion exchange, optics, and photovoltaics.

Microporous Material

A microporous material is a material containing pores with diameters less than 2 nm. Examples of microporous materials include zeolites and metal-organic frameworks. Porous materials are classified into several kinds by their size. According to IUPAC notation, microporous materials have pore diameters of less than 2 nm, mesoporous materials have pore diameters between 2 nm and 50 nm and macroporous materials have pore diameters of greater than 50 nm.

Uses in Laboratories

Microporous materials are often used in laboratory environments to facilitate contaminant-free exchange of gases. Mold spores, bacteria, and other airborne contaminants will become trapped, while allowing gases to pass through the material. This allows for a sterile environment in the contained area.

Use in Medicine

Microporous adhesive tape is a surgical tape used to hold wound dressings and bandages in place, introduced in 1959 by 3M with the trade name *Micropore*. It can be used to hold gauze padding over small wounds, usually as a temporary measure until a suitable dressing is applied. The Steri-Strip was derived from Microporous surgical tape.

Other Uses

Microporous tape is used by some professional extreme yo-yoers to wrap around their fingers and prevent string burn or irritation. Rock climbers use microporous tape to wrap their hands in 'tape gloves', a means of protecting the skin from rock abrasion when jamming hands into cracks as a means of ascent (crack climbing, as opposed to face climbing - gripping holds on the face of the rock).

Microporous media used in large format printing applications normally with a pigment based ink to maintain colour balance and life expectancy of the resultant printed image.

Microporous tape is also used by some film and TV sound recordists to affix small radio microphones to actor's skin.

Characterisation of Pore Space in Soil

Soil is essential to most animals on the earth. It is a relatively thin crust where an even smaller portion contains much of the biological activity. Soil consists of three different phases. A solid phase (H" 50%) that contains mainly minerals of varying sizes as well as organic compounds. The rest is pore space. This space contains the liquid and gas phases. In order to understand porosity better a series of equations have been used to express the quantitative interactions between the three phases of soil.

Macropores or fractures play a major role in infiltration rates in many soils as well as preferential flow patterns, hydraulic conductivity and evapotranspiration. Cracks are also very influential in gas exchange, influencing respiration within soils. Modelling cracks therefore helps understand how these processes work and what the effects of changes in soil cracking such as compaction, can have on these processes.

Bulk Density

$$\rho = \frac{M_s}{t_V}$$

The bulk density of soil depends greatly on the mineral make up of soil and the degree of compaction. The density of quartz is around 2.65 g/cm^3 but the bulk density of a soil may be less than half that density. Most soils have a bulk density between 1.0 and 1.6 g/cm^3 but organic soil and some friable clay may have a bulk density well below 1 g/cm^3. Core samples are taken by driving a metal core into the earth at the desired depth and soil horizon. The samples are then oven dried and weighed.

Bulk density = (mass of oven dry soil)/volume

The bulk density of soil is inversely related to the porosity of the same soil. The more pore space in a soil the lower the value for bulk density.

Porosity (f)

$$f = \frac{V_f}{V_t} \text{ or } f = \frac{V_a + V_w}{V_s + V_a + V_w}$$

Porosity is a measure of the total pore space in the soil. This is measured as a volume or percent. The amount of porosity in a soil depends on the minerals that make up the soil and the amount of sorting that occurs within the soil structure. For example a sandy soil will have larger porosity than silty sand, because the silt will fill in the gaps between the sand particles.

Pore Space Relations

Hydraulic Conductivity

Hydraulic conductivity (K) is a property of soil that describes the ease with which water can move through pore spaces. It depends on the permeability of the material (pores, compaction) and on the degree of saturation. Saturated hydraulic conductivity, K_{sat}, describes water movement through saturated media. Where hydraulic conductivity has the capability to be measured at any state. It can be estimated by numerous kinds of equipment. To calculate hydraulic conductivity, Darcy's law is used. The manipulation of the law depends on the Soil saturation and instrument used.

Infiltration

Infiltration is the process by which water on the ground surface enters the soil. The Water enters the soil through the pores by the forces of gravity and capillary action. The largest cracks and pores offer a great reservoir for the initial flush of water. This allows a rapid infiltration. The smaller pores take longer to fill and rely on capillary forces as well as gravity. The smaller pores have a slower infiltration as the soil becomes more saturated.

Pore Types

A pore is not simply a void in the solid structure of soil. There are three main categories for pore sizes that all have different characteristics and contribute different attributes to soils depending on the number and frequency of each type.

Macropore

The pores that are too large to have any significant capillary force. These pores are full of air at field capacity. Macropores can be caused by cracking, division of peds and aggregates, as well as plant roots, and zoological exploration. Size >75 μm.

Mesopore

The pores filled with water at field capacity. Also known as storage pores because of the ability to store water useful to plants. They do not have capillary forces too great so that the water does not become limiting to the plants. These mesopores are ideally always full or contain liquid to have successful plant growth. The properties of mesopores are highly studied by soil scientists to help with agriculture and irrigation. Size 75 μm–30 μm.

Micropore

The pores that are filled with water at permanent wilting point. These pores are too small for a plant to use without great difficulty. The water associated is usually adsorbed onto the surfaces of clay molecules. The water held in micropores is important to the activity of microbes creating moist anaerobic conditions. The water can also cause either the oxidation or reduction of molecules in the crystalline structure of the soil minerals. Size <30 μm.

Modelling Methods

Basic crack modelling has been undertaken for many years by simple observations and measurements of crack size, distribution, continuity and depth. These observations have either been surface observation or done on profiles in pits. Hand tracing and measurement of crack patterns on paper was one method used prior to advances in modern technology. Another field method was with the use of string and a semicircle of wire. The semi circle was moved along alternating sides of a string line. The cracks within the semicircle were measured for width, length and depth using a ruler. The crack distribution was calculated using the principle of Buffon's needle.

Disc Permeameter

This method relies on the fact that crack sizes have a range of different water potentials. At zero water potential at the soil surface an estimate of saturated hydraulic conductivity is produced, with all pores filled with water. As the potential is decreased progressively larger cracks drain. By measuring at the hydraulic conductivity at a range of negative potentials, the pore size distribution can be determined. While this is not a physical model of the cracks, it does give an indication to the sizes of pores within the soil.

Horgan and Young Model

Horgan and Young (2000) produced a computer model to create a two-dimensional prediction of surface crack formation. It used the fact that once cracks come within a certain distance of one another they tend to be attracted to each other. Cracks also tend to turn within a particular range of angles and at some stage a surface aggregate gets to a size that no more cracking will occur. These are often characteristic of a soil and can therefore be measured in the field and used in the model. However it was not able to predict the points at which cracking starts and although random in the formation of crack pattern, in many ways, cracking of soil is often not random, but follows lines of weaknesses.

Araldite-Impregnation Imaging

A large core sample is collected. This is then impregnated with araldite and a fluorescent resin. The core is then cut back using a grinding implement, very gradually (~1 mm per time), and at every interval the surface of the core sample is digitally imaged. The images are then loaded into a computer where they can be analysed. Depth, continuity, surface area and a number of other measurements can then be made on the cracks within the soil.

Electrical Resistivity Imaging

Using the infinite resistivity of air, the air spaces within a soil can be mapped. A specially designed resistivity metre had improved the metre-soil contact and therefore the area of the reading. This technology can be used to produce images that can be analysed for a range of cracking properties.

Nanoporous Materials

Nanoporous materials consist of a regular organic or inorganic framework supporting a regular, porous structure. The size of the pores is generally 100 nanometres or smaller. Most nanoporous materials can be classified as bulk materials or membranes. Activated carbon and zeolites are two examples of bulk nanoporous materials, while cell membranes can be thought of as nanoporous membranes.

There are many natural nanoporous materials, but artificial materials can also be manufactured. One method of doing so is to combine polymers with different melting points, so that upon heating one polymer degrades. A nanoporous material with consistently sized pores has the property of letting only certain substances pass through, while blocking others.

Mesoporous Silica

Mesoporous silica is a form of silica and a recent development in nanotechnology. The most common types of mesoporous nanoparticles are MCM-41 and SBA-15. Research continues on the particles, which have applications in catalysis, drug delivery and imaging.

A procedure for producing mesoporous silica was patented around 1970. It went almost unnoticed and was reproduced in 1997. Mesoporous silica nanoparticles (MSNs) were independently synthesized in 1990 by researchers in Japan. They were later produced also at Mobil Corporation laboratories and named Mobil Crystalline Materials, or MCM-41. Six years later, silica nanoparticles with much larger 4.6 to 30 nanometre pores were produced at the University of California, Santa Barbara. The material was named Santa Barbara Amorphous type material, or SBA-15. These particles also have a hexagonal array of pores.

The researchers who invented these types of particles planned to use them as molecular sieves. Today, mesoporous silica nanoparticles have many applications in medicine, biosensors, and imaging.

Synthesis

Mesoporous silica nanoparticles are synthesized by reacting tetraethyl orthosilicate with a template made of micellar rods. The result is a collection of nano-sized spheres or rods that are filled with a regular arrangement of pores. The template can then be removed by washing with a solvent adjusted to the proper pH.

In another technique, the mesoporous particle could be synthesized using a simple sol-gel method or a spray drying method. Tetraethyl orthosilicate is also used with an additional polymer monomer (as a template).

Drug Delivery

The large surface area of the pores allows the particles to be filled with a drug or a cytotoxin. Like a Trojan Horse, the particles will be taken up by certain biological cells though endocytosis, depending on what chemicals are attached to the outside of the spheres. Some types of cancer cells will take up more of the particles than healthy cells will, giving researchers hope that MCM-41 will one day be used to treat certain types of cancer.

Ordered mesoporous silica (e.g. SBA-15, TUD-1, HMM-33, and FSM-16) also show potential to boost the in vitro and in vivo dissolution of poorly water soluble drugs. Many drug-candidates coming from drug discovery suffer from a poor water solubility. An insufficient dissolution of these hydrophobic drugs in the gastrointestinal fluids strongly limits the oral bioavailability. One example is itraconazole which is an antimycoticum known for its poor aqueous solubility. Upon introduction of itraconazole-on-SBA-15 formulation in simulated gastrointestinal fluids, a supersaturated solution is obtained giving rise to enhanced transepithelial intestinal transport. Also the efficient uptake into the systemic circulation of SBA-15 formulated itraconazole has been demonstrated in vivo (rabbits and dogs). This approach based on SBA-15 yields stable formulations and can be used for a wide variety of poorly water soluble compounds.

Biosensors

The structure of these particles allows them to be filled with a fluorescent dye that would normally be unable to pass through cell walls. The MSN material is then capped off with a molecule that is compatible with the target cells. When the MSNs are added to a cell culture, they carry the dye across the cell membrane. These particles are optically transparent, so the dye can be seen through the silica walls. The dye in the particles does not have the same problem with self-quenching that a dye in solution has. The types of molecules that are grafted to the outside of the MSNs will control what kinds of biomolecules are allowed inside the particles to interact with the dye.

Silicon Dioxide

Silicon dioxide, also known as silica (from the Latin *silex*), is a chemical compound that is a dioxide of silicon with the chemical formula SiO_2. It has been known since ancient times. Silica is most commonly found in nature as quartz, as well as in various living organisms, Silica is one of the most complex and most abundant families of materials, existing both as several minerals and being produced synthetically. Notable examples include fused quartz, crystal, fumed silica, silica gel, and aerogels. Applications range from structural materials to microelectronics to components used in the food industry.

Silicon dioxide is obtained by mining and purification of the resulting mineral. Quartz comprises more than 10% by mass of the earth's crust. It is also produced, almost always via the processing of quartz, synthetically on a very large scale.

Fumed Silica

Pyrogenic silica (sometimes called fumed silica or silica fume) is a very fine particulate or colloidal form of silicon dioxide. It is prepared by burning $SiCl_4$ in an oxygen rich hydrocarbon flame to produce a "smoke" of SiO_2.

$$SiCl_4 + 2\ H_2 + O_2 \rightarrow SiO_2 + 4\ HCl.$$

Precipitated Silica

Amorphous silica, silica gel, is produced by the acidification of solutions of sodium silicate. An initially formed gelatinous precipitate is then washed and then dehydrated to produce colourless microporous silica. Ideallized equation involving a trisilicate and sulfuric acid is shown:

$$Na_2Si_3O_7 + H_2SO_4 \rightarrow 3\ SiO_2 + Na_2SO_4 + H_2O$$

Approximately one billion kilograms/year (1999) of silica was produced in this manner, mainly for use for polymer composites – tires and shoe soles.

On Microchips

Thin films of silica grow spontaneously on silicon wafers via thermal oxidation. This route gives a very shallow layer (approximately 1 nm or 10 Å) of so-called native oxide. Higher temperatures and alternative environments are used to grow well-controlled layers of silicon dioxide on silicon, for example at temperatures between 600 and 1200 °C, using so-called dry or wet oxidation with O_2 or H_2O, respectively. The depth of the layer of silicon replaced by the dioxide is 44% of the depth of the silicon dioxide layer produced.

The native oxide layer can be beneficial in microelectronics, where it acts as electric insulators with high chemical stability. In electrical applications, it can protect the silicon, store charge, block current, and even act as a controlled pathway to limit current flow.

Laboratory or Speciality Methods

From Silicate Esters

Many routes to silicon dioxide start with silicate esters, the best known being tetraethyl orthosilicate (TEOS). Simply heating TEOS at 680–730 °C gives the dioxide:

$$Si(OC_2H_5)_4 \rightarrow SiO_2 + 2\ O(C_2H_5)_2$$

Similarly TEOS combusts around 400 °C:

$$Si(OC_2H_5)_4 + 12\ O_2 \rightarrow SiO_2 + 10\ H_2O + 8\ CO_2$$

TEOS undergoes hydrolysis via the so-called sol-gel process. The course of the reaction and nature of the product are affected by catalysts, but the idealized equation is:

$$Si(OC_2H_5)_4 + 2\ H_2O \rightarrow SiO_2 + 4\ HOCH_2CH_3$$

Other Methods

Being highly stable, silicon dioxide arises from many methods. Conceptually simple, but of little practical value, combustion of silane give silicon dioxide. This reaction is analogous to the combustion of methane:

$$SiH_4 + 2\ O_2 \rightarrow SiO_2 + 2\ H_2O.$$

Uses

An estimated 95% of silicon dioxide produced is consumed in the construction industry, e.g. for the production of Portland cement. Other major applications are listed below.

Precursor to glass and silicon metal

Silica is used primarily in the production of glass for windows, drinking glasses, beverage bottles, and many other uses. The majority of optical fibres for telecommunication are also made from silica. It is a primary raw material for many ceramics such as earthenware, stoneware, and porcelain. Silicon dioxide is used to produce elemental silicon. The process involves carbothermic reduction in an electric arc furnace:

$$SiO_2 + 2\ C \rightarrow Si + 2\ CO$$

Food and Pharmaceutical Applications

Silica is a common additive in the production of foods, where it is used primarily as a flow agent in powdered foods, or to absorb water in hygroscopic applications. It is the primary component of diatomaceous earth. Colloidal silica is also used as a wine, beer, and juice fining agent.

In pharmaceutical products, silica aids powder flow when tablets are formed.

Other

A silica-based aerogel was used in the Stardust spacecraft to collect extraterrestrial particles. Silica is also used in the extraction of DNA and RNA due to its ability to bind to the nucleic acids under the presence of chaotropes. As hydrophobic silica it is used as a defoamer component. In hydrated form, it is used in toothpaste as a hard abrasive to remove tooth plaque.

In its capacity as a refractory, it is useful in fibre form as a high-temperature thermal protection fabric. In cosmetics, it is useful for its light-diffusing properties and natural absorbency. It is also used as a thermal enhancement compound in ground source heat pump industry.

Structure

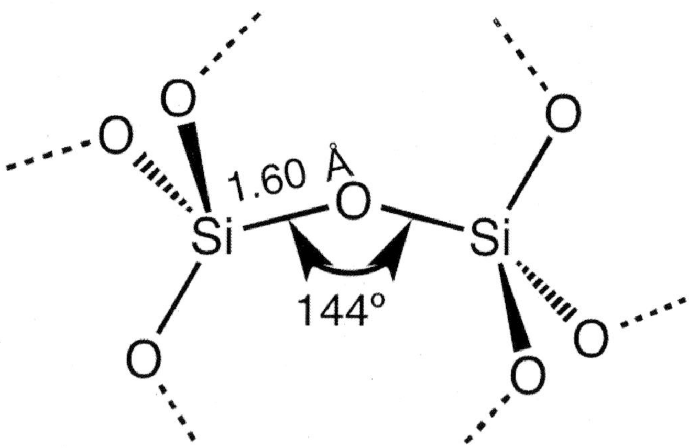

Figure: *Structural motif found in α-quartz, but also found in almost all forms of silicon dioxide.*

In the majority of silicates, the Si atom shows tetrahedral coordination, with 4 oxygen atoms surrounding a central Si atom. The most common example is seen in the quartz crystalline form of silica SiO_2. In each of the most thermodynamically stable crystalline forms of silica, on average, all 4 of the vertices (or oxygen atoms) of the SiO_4 tetrahedra are shared with others, yielding the net chemical formula: SiO_2.

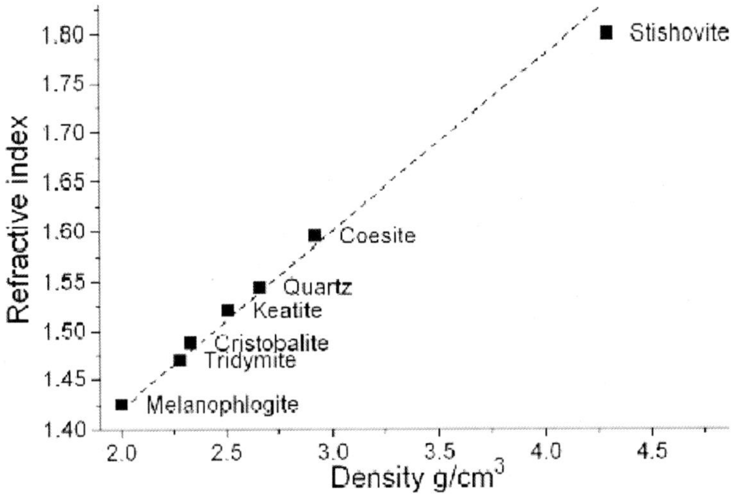

Figure: *Relation between refractive index and density for some SiO_2 forms.*

For example, in the unit cell of α-quartz, the central tetrahedron shares all 4 of its corner O atoms, the 2 face-centred tetrahedra share 2 of their corner O atoms, and the 4 edge-centred tetrahedra share just one of their O atoms with other SiO_4 tetrahedra. This leaves a net average of 12 out of 24 total vertices for that portion of the 7 SiO_4 tetrahedra that are considered to be a part of the unit cell for silica.

SiO_2 has a number of distinct crystalline forms (polymorphs) in addition to amorphous forms. With the exception of stishovite and fibrous silica, all of the crystalline forms involve tetrahedral SiO_4 units linked together by shared vertices in different arrangements.

Silicon–oxygen bond lengths vary between the different crystal forms, for example in α-quartz the bond length is 161 pm, whereas in α-tridymite it is in the range 154–171 pm. The Si-O-Si angle also varies between a low value of 140° in α-tridymite, up to 180° in β-tridymite. In α-quartz the Si-O-Si angle is 144°.

Fibrous silica has a structure similar to that of SiS_2 with chains of edge-sharing SiO_4 tetrahedra. Stishovite, the higher pressure form, in contrast has a rutile like structure where silicon is 6 coordinate. The density of stishovite is 4.287 g/cm^3, which compares to α-quartz, the densest of the low pressure forms, which has a density of 2.648 g/cm^3.

The difference in density can be ascribed to the increase in coordination as the six shortest Si-O bond lengths in stishovite (four Si-O bond lengths of 176 pm and two others of 181 pm) are greater than the Si-O bond length (161 pm) in α-quartz. The change in the coordination increases the ionicity of the Si-O bond.

But more important is the observation that any deviations from these standard parameters constitute microstructural differences or variations, which represent an approach to an amorphous, vitreous or glassy solid.

The only stable form under normal conditions is α-quartz and this is the form in which crystalline silicon dioxide is usually encountered. In nature impurities in crystalline α-quartz can give rise to colours. The high temperature minerals, cristobalite and tridymite, have both a lower density and index of refraction than quartz.

Since the composition is identical, the reason for the discrepancies must be in the increased spacing in the high temperature minerals. As is common with many substances, the higher the temperature the farther apart the atoms due to the increased vibration energy.

The high pressure minerals, seifertite, stishovite, and coesite, on the other hand, have a higher density and index of refraction when compared to quartz. This is probably due to the intense compression of the atoms that must occur during their formation, resulting in a more condensed structure.

Faujasite silica is another form of crystalline silica. It is obtained by dealumination of a low-sodium, ultra-stable Y zeolite with a combined acid and thermal treatment. The resulting product contains over 99% silica, has high crystallinity and high surface area (over 800 m^2/g). Faujasite-silica has very high thermal and acid stability. For example, it maintains a high degree of long-range molecular order (or crystallinity) even after boiling in concentrated hydrochloric acid.

Molten silica exhibits several peculiar physical characteristics that are similar to the ones observed in liquid water: negative temperature expansion, density maximum (at temperatures ~5000 °C), and a heat capacity minimum. Its density decreases from 2.08 g/cm^3 at 1950 °C to 2.03 g/cm^3 at 2200 °C. When molecular silicon monoxide, SiO, is condensed in an argon matrix cooled with helium along with oxygen atoms generated by microwave discharge, molecular SiO_2 is produced with a linear structure.

Dimeric silicon dioxide, $(SiO_2)_2$ has been prepared by reacting O_2 with matrix isolated dimeric silicon monoxide, (Si_2O_2). In dimeric silicon dioxide there are two oxygen atoms bridging between the silicon atoms with an Si-O-Si angle of 94° and bond length of 164.6 pm and the terminal Si-O bond length is 150.2 pm. The Si-O bond length is 148.3 pm, which compares with the length of 161 pm in α-quartz. The bond energy is estimated at 621.7 kJ/mol.

Fused Quartz

When molten silicon dioxide SiO_2 is rapidly cooled, it does not crystallize but solidifies as a glass. The geometry of the silicon and oxygen centres in glass is similar to that in quartz and most other crystalline forms of the same composition, i.e., silicon is surrounded by a regular tetrahedra of oxygen centres.

The difference between the glass and the crystalline forms arise from the connectivity of these tetrahedral units. Although there is no long range periodicity in the glassy network there remains significant ordering at length scales well beyond the SiO bond length. One example of this ordering is found in the preference of the network to form rings of 6-tetrahedra.

The glass transition temperature of pure SiO_2 is about 1475 K.

Chemical Reactions

Silica is converted to silicon by reduction with carbon.

Fluorine reacts with silicon dioxide to form SiF_4 and O_2 whereas the other halogen gases (Cl_2, Br_2, I_2) are essentially unreactive.

Silicon dioxide is attacked by hydrofluoric acid (HF) to produce hexafluorosilicic acid:

$$SiO_2 + 6\ HF \rightarrow H_2SiF_6 + 2\ H_2O.$$

HF is used to remove or pattern silicon dioxide in the semiconductor industry.

Silicon dioxide dissolves in hot concentrated alkali or fused hydroxide, as described in this idealized equation:

$$SiO_2 + 2\ NaOH \rightarrow Na_2SiO_3 + H_2O.$$

Silicon dioxide reacts with basic metal oxides (e.g. sodium oxide, potassium oxide, lead(II) oxide, zinc oxide, or mixtures of oxides forming silicates and glasses as the Si-O-Si bonds in silica are broken successively). As an example the reaction of sodium oxide and SiO_2 can produce sodium orthosilicate, sodium silicate, and glasses, dependent on the proportions of reactants:

$$2\ Na_2O + SiO_2 \rightarrow Na_4SiO_4;$$

$$Na_2O + SiO_2 \rightarrow Na_2SiO_3;$$

$$(0.25\text{–}0.8)Na_2O + SiO_2 \rightarrow \text{glass}.$$

Examples of such glasses have commercial significance, e.g. soda-lime glass, borosilicate glass, lead glass. In these glasses, silica is termed the network former or lattice former.

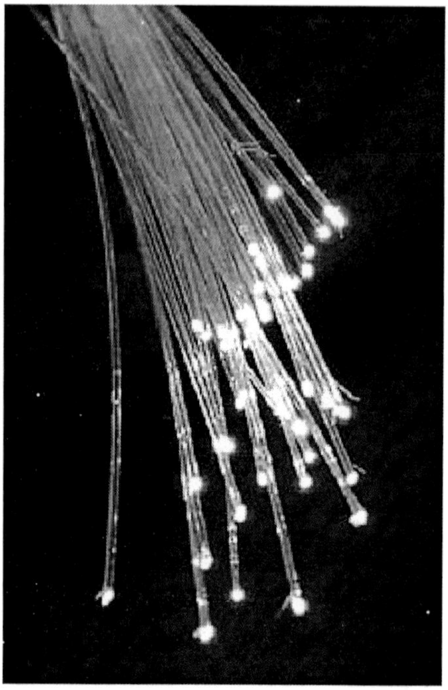

Figure: *Bundle of optical fibres composed of high purity silica.*

Silicon dioxide reacts with elemental silicon at high temperatures to produce SiO:

$$SiO_2 + Si \rightarrow 2\ SiO$$

Solubility in Water

The solubility of silicon dioxide in water strongly depends on its crystalline form and is 3–4 times higher for silica than quartz; as a function of temperature, it peaks at about 340 °C. This property is used to grow single crystals of quartz in a hydrothermal process where natural quartz is dissolved in superheated water in a pressure vessel that is cooler at the top. Crystals of 0.5–1 kg can be grown over a period of 1–2 months. These crystals are a source of very pure quartz for use in electronic applications.

Occurrence

Biology

Even though it is poorly soluble, silica occurs widely in many plants. Plant materials with high silica phytolith content appear to be of importance to grazing animals, from chewing insects to ungulates. Studies have shown that it accelerates tooth wear, and high levels of silica in plants frequently eaten by herbivores may have developed as a defense mechanism against predation. It is also the primary component of rice husk ash, which is used, for example, in filtration and cement manufacturing.

Silicification in and by cells has been common in the biological world for well over a billion years. In the modern world it occurs in bacteria, single-celled organisms, plants, and animals (invertebrates and vertebrates). Prominent examples include:

- Tests or frustules of diatoms and Radiolaria.
- Silica phytoliths in the cells of many plants, including Equisetaceae, practically all grasses, and a wide range of dicotyledons.
- The spicules forming the skeleton of many sponges.

Crystalline minerals formed in the physiological environment often show exceptional physical properties (e.g. strength, hardness, fracture toughness) and tend to form hierarchical structures that exhibit microstructural order over a range of scales. The minerals are crystallized from an environment that is undersaturated with respect to silicon, and under conditions of neutral pH and low temperature (0–40 °C).

Formation of the mineral may occur either within the cell wall of an organism (such as with phytoliths), or outside the cell wall, as typically happens with tests. Specific biochemical reactions exist for mineral deposition. Such reactions include those that involve lipids, proteins, and carbohydrates. It is unclear in what ways silica is important in the nutrition of animal. This field of research is challenging because silica is ubiquitous and in most circumstances dissolves in trace quantities only. All the same it certainly does occur in the living body, leaving us with the problem that it is hard to create proper silica-free controls for purposes of research. This makes it difficult to be sure when the silica present has had operative beneficial effects, and when its presence is coincidental, or even harmful. The current consensus is that it certainly seems important in the growth, strength, and management of many connective tissues. This is true not only for hard connective tissues such as bone and tooth but possibly in the biochemistry of the subcellular enzyme-containing structures as well.

Health Effects

Silica ingested orally is essentially nontoxic, with an LD50 of 5000 mg/kg. On the other hand, inhaling finely divided crystalline silica dust can lead to silicosis, bronchitis, or cancer, as the dust becomes lodged in the lungs and continuously irritates them, reducing lung capacities. Prior to new rules issued in 2013 OSHA allowed 100 micrograms a cubic metre of air. The new regulations reduce the amount to 50 μg/m^3 down from 100 μg/m^3. The exposure limit for the construction industry is also set at 50 μg/m^3 down from 250 μg/m^3.

In the body crystalline silica particles do not dissolve over clinically relevant periods. Silica crystals can activate the NLRP3 inflammasome inside macrophages and dendritic cells and thereby result in processing of pro-Interleukin 1 beta into its mature form. Chronic exposure to silica may thereby account for some of its health hazards, as interleukin-1 is a highly pro-inflammatory cytokine in the immune system. This effect can create an occupational hazard for people working with sandblasting equipment, products that contain powdered crystalline silica and so on. Children, asthmatics of any age, allergy sufferers, and the elderly (all of whom have reduced lung capacity) can be affected in much less time. Amorphous silica, such as fumed silica is not associated with development of silicosis, but may cause irreversible lung damage in some cases. Laws restricting silica exposure *with respect to the silicosis hazard* specify that they are concerned only with silica that is both crystalline and dust-forming.

A study that followed subjects for 15 years found that higher levels of silica in water appeared to decrease the risk of dementia. The study found an association between an increase of 10 milligram-per-day of the intake of silica in drinking water with a decreased risk of dementia of 11%.

Crystalline silica is used in hydraulic fracturing of formation which contain tight oil and shale gas, a use which presents a health hazard to workers. In 2013 OSHA announced tightened restrictions on the amount of crystalline silica which could be present and required "green completion" of fracked wells to reduce exposure during the initial burst of material emitted by a fracked well.

Solar Cell Research

There are currently many research groups active in the field of photovoltaics in universities and research institutions around the world. This research can be divided into three areas: making current technology solar cells cheaper and/or more efficient to effectively compete with other energy sources; developing new technologies based on new solar cell architectural designs; and developing new materials to serve as light absorbers and charge carriers.

Silicon Processing

One way of reducing the cost is to develop cheaper methods of obtaining silicon that is sufficiently pure. Silicon is a very common element, but is normally bound in silica, or silica sand. Processing silica (SiO_2) to produce silicon is a very high energy process - at current efficiencies, it takes one to two years for a conventional solar cell to generate as much energy as was used to make the silicon it contains. More energy efficient methods of synthesis are not only beneficial to the solar industry, but also to industries surrounding silicon technology as a whole.

The current industrial production of silicon is via the reaction between carbon (charcoal) and silica at a temperature around 1700 °C. In this process, known as carbothermic reduction, each tonne of silicon (metallurgical grade, about 98% pure) is produced with the emission of about 1.5 tonnes of carbon dioxide.

Solid silica can be directly converted (reduced) to pure silicon by electrolysis in a molten salt bath at a fairly mild temperature (800 to 900 °C). While this new process is in principle the same as the FFC Cambridge Process which was first discovered in late 1996, the interesting laboratory finding is that such electrolytic silicon is in the form of porous silicon which turns readily into a fine powder, with a particle size of a few micrometres, and may therefore offer new opportunities for development of solar cell technologies.

Another approach is also to reduce the amount of silicon used and thus cost, is by micromachining wafers into very thin, virtually transparent layers that could be used as transparent architectural coverings. The technique involves taking a silicon wafer, typically 1 to 2 mm thick, and making a multitude of parallel, transverse slices across the wafer, creating a large number of slivers that have a thickness of 50 micrometres and a width equal to the thickness of the original wafer. These slices are rotated 90 degrees, so that the surfaces corresponding to the faces of the original wafer become the edges of the slivers. The result is to convert, for example, a 150 mm diameter, 2 mm-thick wafer having an exposed silicon surface area of about 175 cm^2 per side into about 1000 slivers having dimensions of 100 mm × 2 mm × 0.1 mm, yielding a total exposed silicon surface area of about 2000 cm^2 per side. As a result of this rotation, the electrical doping and contacts that were on the face of the wafer are located at the edges of the sliver, rather than at the front and rear as in the case of conventional wafer cells. This has the interesting effect of making the cell sensitive from both the front and rear of the cell (a property known as bifaciality). Using this technique, one silicon wafer is enough to build a 140 watt panel, compared to about 60 wafers needed for conventional modules of same power output.

Nanocrystalline Solar Cells

These structures make use of some of the same thin-film light absorbing materials but are overlain as an extremely thin absorber on a supporting matrix of conductive polymer or mesoporous metal oxide having a very high surface area to increase internal reflections (and hence increase the probability of light absorption). Using nanocrystals allows one to design architectures on the length scale of nanometres, the typical exciton diffusion length. In particular, single-nanocrystal ('channel') devices, an array of single p-n junctions between the electrodes and separated by a period of about a diffusion length, represent a new architecture for solar cells and potentially high efficiency.

Thin-Film Processing

Thin-film photovoltaic cells can use less than 1% of the expensive raw material (silicon or other light absorbers) compared to wafer-based solar cells, leading to a significant price drop per Watt peak capacity. There are many research groups around the world actively researching different thin-film approaches and/or materials.

One particularly promising technology is crystalline silicon thin films on glass substrates. This technology combines the advantages of crystalline silicon as a solar cell material (abundance, non-toxicity, high efficiency, long-term stability) with the cost savings of using a thin-film approach.

Another interesting aspect of thin-film solar cells is the possibility to deposit the cells on all kind of materials, including flexible substrates (PET for example), which opens a new dimension for new applications.

Metamorphic Multijunction Solar Cell

The National Renewable Energy Laboratory (NREL) won one of *R&D Magazine*'s R&D 100 Awards for its Metamorphic Multijunction photovoltaic cell, an ultra-light and flexible cell that converts solar energy with record efficiency.

The ultra-light, highly efficient solar cell was developed at NREL and is being commercialized by Emcore Corp. of Albuquerque, N.M., in partnership with the Air Force Research Laboratories Space Vehicles Directorate at Kirtland Air Force Base in Albuquerque.

It represents a new class of solar cells with clear advantages in performance, engineering design, operation and cost. For decades, conventional cells have featured wafers of semiconducting materials with similar crystalline structure. Their performance and cost effectiveness is constrained by growing the cells in an upright configuration. Meanwhile, the cells are rigid, heavy and thick with a bottom layer made of germanium.

In the new method, the cell is grown upside down. These layers use high-energy materials with extremely high quality crystals, especially in the upper layers of the cell where most of the power is produced. Not all of the layers follow the lattice pattern of even atomic spacing. Instead, the cell includes a full range of atomic spacing, which allows for greater absorption and use of sunlight.

The thick, rigid germanium layer is removed, reducing the cell's cost and 94% of its weight. By turning the conventional approach to cells on its head, the result is an ultra-light and flexible cell that also converts solar energy with record efficiency (40.8% under 326 suns concentration).

Polymer Processing

The invention of conductive polymers (for which Alan Heeger, Alan G. MacDiarmid and Hideki Shirakawa were awarded a Nobel prize) may lead to the development of much cheaper cells that are based on inexpensive plastics. However, organic solar cells generally suffer from degradation upon exposure to UV light, and hence have lifetimes which are far too short to be viable.

The bonds in the polymers, are always susceptible to breaking up when radiated with shorter wavelengths. Additionally, the conjugated double bond systems in the polymers which carry the charge, react more readily with light and oxygen. So most conductive polymers, being highly unsaturated and reactive, are highly sensitive to atmospheric moisture and oxidation, making commercial applications difficult.

Nanoparticle Processing

Experimental non-silicon solar panels can be made of quantum heterostructures, e.g. carbon nanotubes or quantum dots, embedded in conductive polymers or mesoporous metal oxides. In addition, thin films of many of these materials on conventional silicon solar cells can increase the optical coupling efficiency into the silicon cell, thus boosting the overall efficiency. By varying the size of the quantum dots, the cells can be tuned to absorb different wavelengths. Although the research is still in its infancy, quantum dot modified photovoltaics may be able to achieve up to 42% energy conversion efficiency due to multiple exciton generation (MEG).

MIT researchers have found a way of using a virus to improve solar cell efficiency by a third.

Transparent Conductors

Many new solar cells use transparent thin films that are also conductors of electrical charge. The dominant conductive thin films used in research now are transparent conductive oxides (abbreviated "TCO"), and include fluorine-doped tin oxide (SnO_2:F, or "FTO"), doped zinc oxide (e.g.: ZnO:Al), and indium tin oxide (abbreviated "ITO"). These conductive films are also used in the LCD industry for flat panel displays. The dual function of a TCO allows light to pass through a substrate window to the active light-absorbing material beneath, and also serves as an ohmic contact to transport photogenerated charge carriers away from that light-absorbing material. The present TCO materials are effective for research, but perhaps are not yet optimized for large-scale photovoltaic production. They require very special deposition conditions at high vacuum, they can sometimes suffer from poor mechanical strength, and most have poor transmittance in the infrared portion of the spectrum (e.g.: ITO thin films can also be used as infrared filters in airplane windows). These factors make large-scale manufacturing more costly.

A relatively new area has emerged using carbon nanotube networks as a transparent conductor for organic solar cells. Nanotube networks are flexible and can be deposited on surfaces a variety of ways. With some treatment, nanotube films can be highly transparent in the infrared, possibly enabling efficient low-bandgap solar cells. Nanotube networks are p-type conductors, whereas traditional transparent conductors are exclusively n-type. The availability of a p-type transparent conductor could lead to new cell designs that simplify manufacturing and improve efficiency.

Silicon Wafer-Based Solar Cells

Despite the numerous attempts at making better solar cells by using new and exotic materials, the reality is that the photovoltaics market is still dominated by silicon wafer-based solar cells (first-generation solar cells). This means that most solar cell manufacturers are currently equipped to produce this type of solar cells. Consequently, a large body of research is being done all over the world to manufacture silicon wafer-based solar cells at lower cost and to increase the conversion efficiencies without an exorbitant increase in production cost. The ultimate goal for both wafer-based and alternative photovoltaic concepts is to produce solar electricity at a cost comparable to currently market-dominant coal, natural gas, and nuclear power in order to make it the leading primary energy source. To achieve this it may be necessary to reduce the cost of installed solar systems from currently about US$ 1.80

(for bulk Si technologies) to about US$ 0.50 per Watt peak power. Since a major part of the final cost of a traditional bulk silicon module is related to the high cost of solar grade polysilicon feedstock (about US$ 0.4/Watt peak) there exists substantial drive to make Si solar cells thinner (material savings) or to make solar cells from cheaper upgraded metallurgical silicon (so called "dirty Si").

IBM has a semiconductor wafer reclamation process that uses a specialised pattern removal technique to repurpose scrap semiconductor wafers to a form used to manufacture silicon-based solar panels. The new process was recently awarded the "2007 Most Valuable Pollution Prevention Award" from The National Pollution Prevention Roundtable (NPPR).

Infrared Solar Cells

Researchers at Idaho National Laboratory, along with partners at Lightwave Power Inc. in Cambridge, MA and Patrick Pinhero of the University of Missouri, have devised an inexpensive way to produce plastic sheets containing billions of nanoantennas that collect heat energy generated by the sun and other sources, which garnered two 2007 Nano50 awards. The company ceased operations in 2010. While methods to convert the energy into usable electricity still need to be developed, the sheets could one day be manufactured as lightweight "skins" that power everything from hybrid cars to computers and iPods with higher efficiency than traditional solar cells. The nanoantennas target mid-infrared rays, which the Earth continuously radiates as heat after absorbing energy from the sun during the day; also double-sided nanoantenna sheets can harvest energy from different parts of the Sun's spectrum. In contrast, traditional solar cells can only use visible light, rendering them idle after dark.

UV Solar Cells

Japan's National Institute of Advanced Industrial Science and Technology (AIST) has succeeded in developing a transparent solar cell that uses ultraviolet (UV) light to generate electricity but allows visible light to pass through it. Most conventional solar cells use visible and infrared light to generate electricity. Used to replace conventional window glass, the installation surface area could be large, leading to potential uses that take advantage of the combined functions of power generation, lighting and temperature control.

This transparent, UV-absorbing system was achieved by using an organic-inorganic heterostructure made of the p-type semiconducting polymer PEDOT:PSS film deposited on a Nb-doped strontium titanate substrate. PEDOT:PSS is easily fabricated into thin films due to its stability in air and its solubility in water. These solar cells are only activated in the UV region and result in a relatively high quantum yield of 16% electron/photon. Future work in this technology involves replacing the strontium titanate substrate with a strontium titanate film deposited on a glass substrate in order to achieve a low-cost, large-area manufacture.

Since then, other methods have been discovered to include the UV wavelengths in solar cell power generation. Some companies report using nano-phosphors as a transparent coating to turn UV light into visible light. Others have reported extending the absorption range of single-junction photovoltaic cells by doping a wide band gap transparent semiconductor such as GaN with a transition metal such as manganese.

Flexible Solar Cell Research

Flexible solar cell research is a research-level technology, an example of which was created at the Massachusetts Institute of Technology in which solar cells are manufactured by depositing photovoltaic material on flexible substrates, such as ordinary paper, using chemical vapor deposition technology. The technology for manufacturing solar cells on paper was developed by a group of researchers from the Massachusetts Institute of Technology with support from the National Science Foundation and the Eni-MIT Alliance Solar Frontiers Program.

3D Solar Cells

Three-dimensional solar cells that capture nearly all of the light that strikes them and could boost the efficiency of photovoltaic systems while reducing their size, weight and mechanical complexity. The new 3D solar cells, created at the Georgia Tech Research Institute, capture photons from sunlight using an array of miniature "tower" structures that resemble high-rise buildings in a city street grid.

Luminescent Solar Concentrator

Luminescent solar concentrators convert sunlight or other sources of light into preferred frequencies; they concentrate the output for conversion into desirable forms of power, such as electricity. They rely on luminescence, typically fluorescence, in media such as liquids, glasses, or plastics treated with a suitable coating or dopant. The structures are configured to direct the output from a large input area onto a small converter, where the concentrated energy generates photoelectricity. The objective is to collect light over a large area at low cost; luminescent concentrator panels can be made cheaply from materials such as glasses or plastics, while photovoltaic cells are high-precision, high-technology devices, and accordingly expensive to construct in large sizes.

Research is in progress at universities such as Radboud University Nijmegen and Delft University of Technology. For example at Massachusetts Institute of Technology researchers have developed approaches for conversion of windows into sunlight concentrators for generation of electricity. They paint a mixture of dyes onto a pane of glass or plastic. The dyes absorb sunlight and re-emit it as fluorescence within the glass, where it is confined by internal reflection, emerging at the edges of the glass, where it encounters solar cells optimized for conversion of such concentrated sunlight. The concentration factor is about 40, and the optical design yields a solar concentrator that unlike lens-based concentrators, need not be directed accurately at the sun, and can produce output even from diffuse light. Covalent Solar is working on commercialization of the process.

Metamaterials

Metamaterials are heterogeneous materials employing the juxtaposition of many microscopic elements, giving rise to properties not seen in ordinary solids. Using these, it *may* become possible to fashion solar cells that are excellent absorbers over a narrow range of wavelengths. High absorption in the microwave regime has been demonstrated, but not yet in the 300-1100-nm wavelength regime.

Photovoltaic Thermal Hybrid

Photovoltaic thermal hybrid solar collectors, sometimes known as hybrid PV/T systems or PVT, are systems that convert solar radiation into thermal and electrical energy. These systems combine a photovoltaic cell, which converts electromagnetic radiation (photons) into electricity, with a solar thermal collector, which captures the remaining energy and removes waste heat from the PV module. The capture of both electricity and heat allow these devices to have higher exergy and thus be more overall energy efficient than solar photovoltaic (PV) or solar thermal alone. A significant amount of research has gone into developing PVT technology since the 1970s.

Photovoltaic cells suffer from a drop in efficiency with the rise in temperature due to increased resistance. Such systems can be engineered to carry heat away from the PV cells thereby cooling the cells and thus improving their efficiency by lowering resistance. Although this is an effective method, it causes the thermal component to under-perform compared to a solar thermal collector. Recent research showed that photovoltaic materials with low temperature coefficients such as amorphous silicon (a-Si:H) PV allow the PVT to be operated at high temperatures, creating a more symbiotic PVT system. This advantage can be tuned by controlling the dispatch strategy of thermal annealing cycles.

Some systems combine photovoltaic with thermal solar, with the advantage that the thermal solar part carries heat away and cools the photovoltaic cells. Keeping temperature down lowers the resistance and improves the cell efficiency.

System Types

A number of PV/T collectors in different categories are commercially available and can be divided into the following categories:

- PV/T liquid collector
- PV/T air collector
- PV/Ta Liquid and air collector
- PV/T concentrator (CPVT)

PV/T Liquid Collector

The basic water-cooled design uses conductive-metal piping or plates attached to the back of a PV module. The fluid flow arrangement through the cooling element will determine which systems the panels are most suited to.

In a standard fluid based system, a working fluid, typically water, glycol or mineral oil is then piped through these pipes or plate chillers. The heat from the PV cells are conducted through the metal and absorbed by the working fluid (presuming that the working fluid is cooler than the operating temperature of the cells).

In closed-loop systems this heat is either exhausted (to cool it), or transferred at a heat exchanger, where it flows to its application. In open-loop systems, this heat is used, or exhausted before the fluid returns to the PV cells.

It is also possible to disperse nanoparticles in the liquid to create a liquid filter for PV/T applications. The basic advantage of this type of split configuration is that the thermal collector and the photovoltaic collector can operate at different temperatures.

PV/T Concentrator (CPVT)

A concentrating system has the advantage to reduce the amount of solar cells needed. It also can get very good solar thermal performance compared to flat PV/T collectors. The main obstacles are to provide good cooling of the solar cells and a durable tracking system.

Aluminosilicate / PI-b-PEO for Mesoporous Aluminosilicates

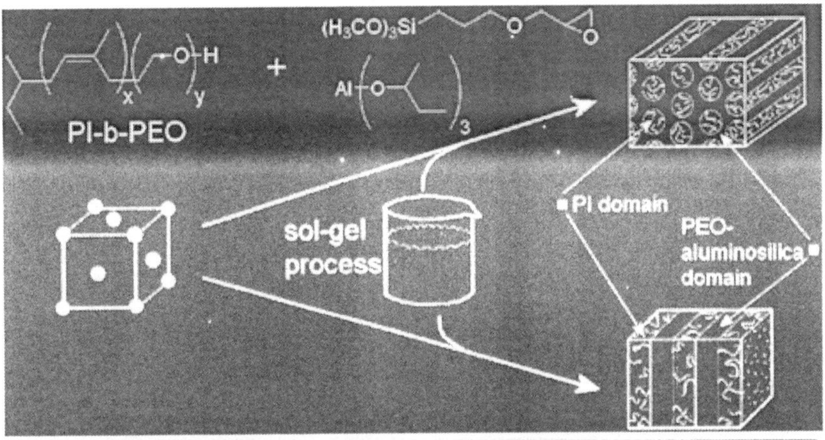

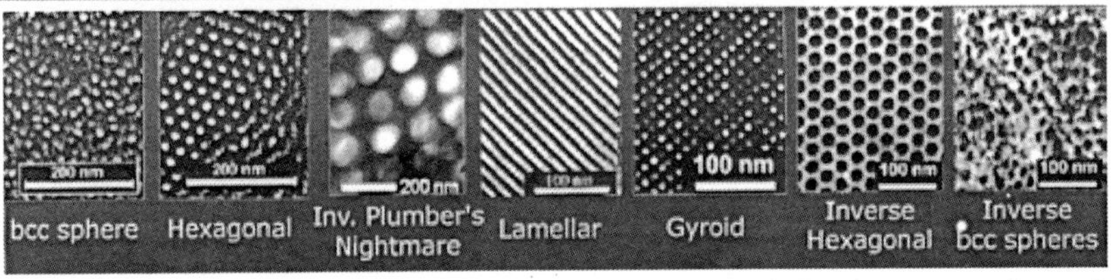

In one example, aluminosilicate nanoparticles are produced in a sol-gel process, where an organically modified silicon alkoxide and an aluminum alkoxide are mixed with water. Slowly drying the solution of nanoparticles and the block copolymer, PI-b-PEO produces well-ordered mesostructures with domains consisting of PI (hydrophobic) and PEO+Aluminosilicate (hydrophilic). By varying the amount of aluminosilicate nanoparticles in the hybrids, a wide range of morphologies can be produced which, in effect, map out the phase diagram of a typical block copolymer.

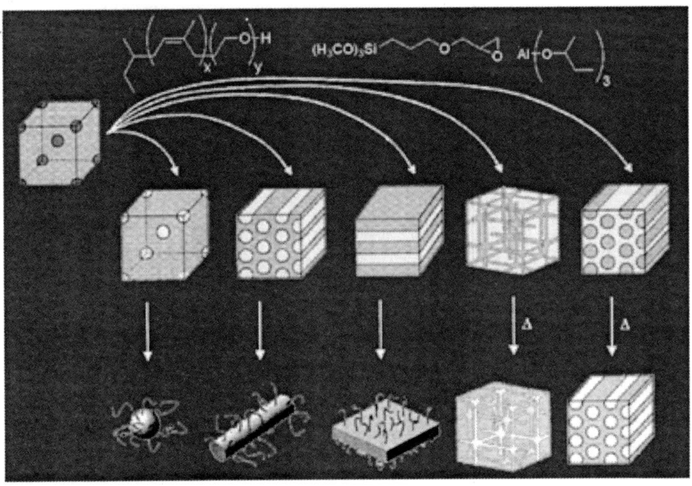

The morphologies typically obtained are (from left to right below) bcc spheres, hexagonally packed cylinders, inverse plumber's nightmare (cubic bicontinuous) lamellae, gyroid, inverse hexagonal cylinders and inverse bcc spheres. These different structures are produced by adding increasing volume fractions of the aluminosilicate-type nanoparticles.

These aluminosilicate-PI-b-PEO hybrids can then either be dissolved to produce "hairy" nano-objects such as spheres, cylinders and lamellae, or heat-treated to burn out the organic material leaving behind mesostructured aluminosilicate ceramic materials. Mesoporous transition metal oxides that have potential applications in energy-harvesting materials have been synthesized in a similar sol-gel route by using PI-b-PEO as structure-directing materials. These materials were crystallized after carbon scaffold, which originated from polyisoprene reinforced the walls of the mesopores, thereby avoiding significant structure collapse (Combined Assemblies of Soft and Hard chemistries, CASH method).

Triblock terpolymers such as poly(isoprene-b-styrene-b-ethylene oxide) have also been used for structure-directing niobia. We find that the phase spaces for 3D network morphologies with triblock terpolymers are larger compared to those of diblock copolymers, giving us more tunability on the hybrid design.

A more exotic four-layer woodpile lattice structure was synthesized via coassembly with a poly(ethylene-co-propylene)-b-(ethylene oxide)-b-(hexyl methacrylate) triblock terpolymer where the middle block is swollen with aluminosilicate nanoparticles.

7

Bi2(Te, Se)3 Nanowire Array Composites as Thermal Energy Harvesters

Bismuth telluride (Bi_2Te_3) is a gray powder that is a compound of bismuth and tellurium also known as bismuth(III) telluride. It is a semiconductor which, when alloyed with antimony or selenium is an efficient thermoelectric material for refrigeration or portable power generation. Topologically protected surface states have been observed in bismuth telluride.

Thermoelectric Properties

Bismuth telluride is a narrow gap layered semiconductor with a trigonal unit cell. The valence and conduction band structure can be described as a many-ellipsoidal model with 6 constant-energy ellipsoids that are centred on the reflection planes. Bi_2Te_3 cleaves easily along the trigonal axis due to Van der Waals bonding between neighbouring tellurium atoms. Due to this, bismuth telluride based material that are used for power generation or cooling applications must be polycrystalline. Furthermore, the Seebeck coefficient of bulk Bi_2Te_3 becomes compensated around room temperature, forcing the materials used in power generation devices to be an alloy of bismuth, antimony, tellurium, and selenium.

Recently, researchers have attempted to improve the efficiency of Bi_2Te_3 based materials by creating structures where one or more dimensions are reduced, such as nanowires or thin films. In one such instance n-type bismuth telluride was shown to have an improved Seebeck coefficient (voltage per unit temperature difference) of "287 μV/K at 54 Celsius. However, one must realise that Seebeck Coefficient and electrical conductivity have a tradeoff; a higher Seebeck coefficient results in decreased carrier concentration and decreased electrical conductivity.

In another case, researchers report that bismuth telluride has high electrical conductivity of 1.1×10^5 S·m/m^2 with its very low lattice thermal conductivity of 1.20 W/(m·K), similar to ordinary glass.

Occurrence

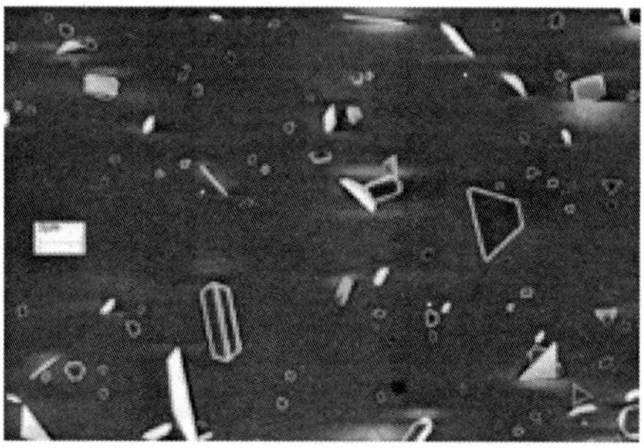

Figure: *A scanning electron microscopy image of small crystals of bismuth telluride, in the same form as tellurobismuthite*

The mineral form of Bi_2Te_3 is tellurobismuthite which is moderately rare. There are many natural bismuth tellurides of different stoichiometry, as well as compounds of the Bi-Te-S-(Se) system, like Bi_2Te_2S (tetradymite).

Preparation

Bismuth Telluride is prepared by sealing a sample of bismuth and tellurium metal in a quartz tube under vacuum (critical, as an unsealed or leaking sample may explode in a furnace) and heating it to 800°C in a muffle furnace.

Thermoelectric Materials

Thermoelectric materials show the thermoelectric effect in a strong or convenient form. The *thermoelectric effect* refers to phenomena by which either a temperature difference creates an electric potential or an electric potential creates a temperature difference. These phenomena are known more specifically as the Seebeck effect (converting temperature to current), Peltier effect (converting current to temperature), and Thomson effect (conductor heating/cooling). While all materials have a nonzero thermoelectric effect, in most materials it is too small to be useful. However, low-cost materials that have a sufficiently strong thermoelectric effect (and other required properties) could be used in applications including power generation and refrigeration.

A commonly used thermoelectric material in such applications is bismuth telluride (Bi2Te3).

Applications

Power Generation

Approximately 90% of the world's electricity is generated by heat energy, typically operating at 30–40% efficiency, losing roughly 15 terawatts of power in the form of heat to the environment. Thermoelectric devices could convert some of this waste heat into useful electricity. Thermoelectric efficiency depends on the *figure of merit*, ZT. There is no theoretical upper limit to ZT, and as ZT approaches infinity, the thermoelectric efficiency approaches the Carnot limit. However, no known thermoelectrics have a ZT>3. As of 2010,, thermoelectric generators serve application niches where efficiency and cost are less important than reliability, light weight, and small size.

Internal combustion engines capture 20–25% of the energy released during fuel combustion. Increasing the conversion rate can increase mileage and provide more electricity for on-board controls and creature comforts (stability controls, telematics, navigation systems, electronic braking, etc.) It may be possible to shift energy draw from the engine (in certain cases) to the electrical load in the car, e.g. electrical power steering or electrical coolant pump operation.

Cogeneration power plants use the heat produced during electricity generation for alternative purposes. Thermoelectrics may find applications in such systems or in solar thermal energy generation.

Refrigeration

Thermoelectric materials can be used as refrigerators, called "thermoelectric coolers", or "Peltier coolers" after the Peltier effect that controls their operation. As a refrigeration technology, Peltier cooling is far less common than vapor-compression refrigeration. The main advantages of a Peltier cooler (compared to a vapor-compression refrigerator) are its lack of moving parts or circulating fluid, and its small size and flexible shape (form factor). Another advantage is that Peltier coolers do not require refrigerant fluids, such as chlorofluorocarbons (CFCs) and related chemicals, which can have harmful environmental effects.

The main *disadvantage* of Peltier coolers is that they cannot simultaneously have low cost and high power efficiency. Advances in thermoelectric materials may allow the creation of Peltier coolers that are both cheap and efficient. It is estimated that materials with ZT>3 (about 20–30% Carnot efficiency) are required to replace traditional coolers in most applications. Today, Peltier coolers are only used in niche applications.

Materials Selection Criteria

Power Factor

The Seebeck coefficient is not the only number that determines the usefulness of a material in a thermoelectric generator or a thermoelectric cooler. Under a given temperature difference, the ability of a material to produce useful electrical power is quantified by its power factor,

$$\text{Power factor} = \sigma S^2.$$

where S is the Seebeck coefficient, and σ is the electrical conductivity. Materials with high power factor are able to generate more energy in a space-constrained application, but they are not necessarily efficient.

Device Efficiency

The efficiency of a thermoelectric device for electricity generation is given by η, defined as

$$\eta = \frac{\text{energy provided to the load}}{\text{heat energy absorbed at hot junction}}.$$

The ability of a given material to efficiently produce thermoelectric power is related to its dimensionless figure of merit given by:

$$ZT = \frac{\sigma S^2 T}{\lambda},$$

which depends on the Seebeck coefficient S, thermal conductivity λ,0 and electrical conductivity σ, and temperature T.

In an actual thermoelectric device, two materials are used. The maximum efficiency η_{max} is then given by

$$\eta_{max} = \frac{T_H - T_C}{T_H} \frac{\sqrt{1+Z\bar{T}} - 1}{\sqrt{1+Z\bar{T}} + \frac{T_C}{T_H}},$$

where T_H is the temperature at the hot junction and T_C is the temperature at the surface being cooled. $Z\bar{T}$ is the modified dimensionless figure of merit, which takes into consideration the thermoelectric capacity of both thermoelectric materials being used in the device and, after geometrical optimization regarding the legs sections, is defined as

$$Z\overline{T} = \frac{(S_p - S_n)^2 \overline{T}}{[(\rho_n \kappa_n)^{1/2} + (\rho_p \kappa_p)^{1/2}]^2}$$

where ρ is the electrical resistivity, $\overline{T}$ is the average temperature between the hot and cold surfaces and the subscripts n and p denote properties related to the n- and p-type semiconducting thermoelectric materials, respectively. Since thermoelectric devices are heat engines, their efficiency is limited by the Carnot efficiency, hence the T_H and T_C terms in η_{max}. Regardless, the coefficient of performance of current commercial thermoelectric refrigerators ranges from 0.3 to 0.6, one-sixth the value of traditional vapor-compression refrigerators.

Phonon-Glass, Electron-Crystal Behaviour

In the efficiency equations above, thermal conductivity and electrical conductivity compete. G. A. Slack proposed that in order to optimize the figure of merit, phonons, which are responsible for thermal conductivity must experience the material as they would in a glass (experiencing a high degree of phonon scattering—lowering thermal conductivity) while electrons must experience it as a crystal (experiencing very little scattering—maintaining electrical conductivity). The figure of merit can be improved through the independent adjustment of these properties.

Semiconductors

Semiconductors are ideal thermoelectric devices because of their band structure and electronic properties at high temperatures. Device efficiency is proportional to *ZT*, so ideal materials have a large Z value at high temperatures. Since temperature is easily adjustable, electrical conductivity is crucial. Specifically, maximizing electrical conductivity at high temperatures and minimizing thermal conductivity optimizes ZT.

Thermal Conductivity

$$\kappa = \kappa_{electron} + \kappa_{phonon}$$

According to the Wiedemann–Franz law, the higher the electrical conductivity, the higher $\kappa_{electron}$ becomes. Therefore, it is necessary to minimize κ_{phonon}. In semiconductors, $\kappa_{electron} < \kappa_{phonon}$, so it is easier to decouple κ and σ in a semiconductor through engineering κ_{phonon}.

Electrical Conductivity

Metals are typically good electrical conductors, but the higher the temperature, the lower the conductivity. This tendency can be explained (approximately) in terms of the Drude conductivity formula:

$\sigma = ne^2\tau/m$

- n is charge carrier density
- e is charge per carrier (elementary charge)
- τ is carrier mean free time between scattering events
- m is carrier mass

As temperature increases, τ decreases while the other numbers stay constant, thereby decreasing σ_{metal}.

In contrast, the electrical conductivity of a semiconductors generally increases with temperature. In semiconductors, carrier mean free time decreases with increasing temperature, however carrier density increases faster with increasing temperature, resulting in increasing $\sigma_{semiconductor}$.

State Density

The band structure of semiconductors offers better thermoelectric effects than the band structure of metals.

The Fermi energy is below the conduction band causing the state density to be asymmetric around the Fermi energy. Therefore, the average electron energy is higher than the Fermi energy, making the system conducive for charge motion into a lower energy state.

By contrast, the Fermi energy lies in the conduction band in metals. This makes the state density symmetric about the Fermi energy so that the average conduction electron energy is close to the Fermi energy, reducing the forces pushing for charge transport. Therefore, semiconductors are ideal thermoelectric materials.

Materials of Interest

Strategies to improve thermoelectrics include both advanced bulk materials and the use of low-dimensional systems. Such approaches to reduce lattice thermal conductivity fall under three general material types:

(1) Alloys: create point defects, vacancies, or rattling structures (heavy-ion species with large vibrational amplitudes contained within partially filled structural sites) to scatter phonons within the unit cell crystal.

(2) Complex crystals: separate the phonon-glass from the electron crystal using approaches similar to those for superconductors. The region responsible for electron transport would be an electron-crystal of a high-mobility semiconductor, while the phonon-glass would be ideal to house disordered structures and dopants without

disrupting the electron-crystal (analogous to the charge reservoir in high-T_c superconductors.)

(3) Multiphase nanocomposites: scatter phonons at the interfaces of nanostructured materials, be they mixed composites or thin film superlattices.

Materials under consideration for thermoelectric device applications include:

Bismuth Chalcogenides

Materials such as Bi2Te3 and Bi2Se3 comprise some of the best performing room temperature thermoelectrics with a temperature-independent thermoelectric effect, ZT, between 0.8 and 1.0. Nanostructuring these materials to produce a layered superlattice structure of alternating Bi2Te3 and Bi2Se3 layers produces a device within which there is good electrical conductivity but perpendicular to which thermal conductivity is poor.

The result is an enhanced ZT (approximately 2.4 at room temperature for p-type). Note that this high value has not entirely been independently confirmed.

Bismuth telluride and its solid solutions are good thermoelectric materials at room temperature and therefore suitable for refrigeration applications around 300 K. The Czochralski method has been used to grow single crystalline bismuth telluride compounds. These compounds are usually obtained with directional solidification from melt or powder metallurgy processes.

Materials produced with these methods have lower efficiency than single crystalline ones due to the random orientation of crystal grains, but their mechanical properties are superior and the sensitivity to structural defects and impurities is lower due to high optimal carrier concentration.

The required carrier concentration is obtained by choosing a nonstoichiometric composition, which is achieved by introducing excess bismuth or tellurium atoms to primary melt or by dopant impurities. Some possible dopants are halogens and group IV and V atoms. Due to the small bandgap (0.16 eV) Bi_2Te_3 is partially degenerate and the corresponding Fermi-level should be close to the conduction band minimum at room temperature. The size of the band-gap means that Bi_2Te_3 has high intrinsic carrier concentration. Therefore, minority carrier conduction cannot be neglected for small stoichiometric deviations. Use of telluride compounds is limited by the toxicity and rarity of tellurium.

Lead Telluride

In 2008 Joseph Heremans and his colleagues have demonstrated that with thallium-doped lead telluride alloy (PbTe) it is possible to achieve ZT of 1.5 at 773 K. Later, Snyder and his colleagues reported ZT~1.4 at 750 K in sodium-doped PbTe, and ZT~1.8 at 850 K in sodium-doped $PbTe_{1-x}Se_x$ alloy. Snyder's group has determined that both thallium and sodium alter the electronic structure of the crystal increasing electric conductivity. They also claim that selenium increases electric conductivity and reduces thermal conductivity.

Inorganic Clathrates

Inorganic clathrates have a general formula $A_xB_yC_{46-y}$ (type I) and $A_xB_yC_{136-y}$ (type II), in these formulas B and C are group III and IV atoms, respectively, which form the framework where "guest" atoms A (alkali or alkaline earth metal) are encapsulated in two different polyhedra facing each other. The differences between types I and II comes from number and size of voids present in their unit cells. Transport properties depend a lot on the properties of the framework, but tuning is possible through the "guest" atoms.

The most direct approach to the synthesis and optimization of thermoelectric properties of semiconducting type I clathrates is substitutional doping, where some framework atoms are replaced with dopant atoms. In addition, powder metallurgical and crystal growth techniques have been used in the synthesis of clathrates. The structural and chemical properties of clathrates enable the optimization of their transport properties with stoichiometry. Type II materials should be investigated in future because their structure allows a partial filling of the polyhedron enabling a better tuning of the electrical properties and therefore a better control of the doping level. Partially filled variants can also be synthesized as semiconducting or even insulating.

Blake *et al.* have predicted ZT~0.5 at room temperature and ZT~1.7 at 800 K for optimized compositions. Kuznetsov *et al.* measured electrical resistance and Seebeck coefficient for three different type I clathrates above room temperature and by estimating high temperature thermal conductivity from the published low temperature data they obtained ZT~0.7 at 700 K for $Ba_8Ga_{16}Ge_{30}$ and ZT~0.87 at 870 K for $Ba_8Ga_{16}Si_{30}$.

Magnesium Group IV Compounds

Mg_2B^{IV} (B^{IV}=Si, Ge, Sn) compounds and their solid solutions are good thermoelectric materials and their figure of merit values are comparable with those of established materials. Due to a lack of systematic studies about their thermoelectric properties, however, the suitability of these materials, and in particular their quasi-ternary solutions, for thermoelectric energy conversion remains in question. The appropriate production methods are based on direct co-melting, but mechanical alloying has also been used. During synthesis, magnesium losses due to evaporation and segregation of components (especially for Mg_2Sn) need to be taken into account. Directed crystallization methods can produce single crystalline material. Solid solutions and doped compounds have to be annealed in order to produce homogeneous samples. At 800 K $Mg_2Si_{1-x}Sn_x$ has been reported to have a figure of merit about 0.9.

Silicides

Higher silicides seem promising materials for thermoelectric energy conversion, because their figure of merit is at the level with materials currently in use and they are mechanically and chemically strong and therefore can often be used in harsh environments without any protection. More detailed studies are needed to assess their potential in thermoelectrics and possibly to find a way to increase their figure of merit. Some of possible fabrication methods are Czochralski and floating zone for single crystals and hot pressing and sintering for polycrystalline.

Skutterudite Thermoelectrics

Recently, skutterudite materials have sparked the interest of researchers in search of new thermoelectrics. These structures are of the form (Co,Ni,Fe)(P,Sb,As).3 and are cubic with space group Im3. Unfilled, these materials contain voids into which low-coordination ions (usually rare earth elements) can be inserted in order to alter thermal conductivity by producing sources for lattice phonon scattering and decrease thermal conductivity due to the lattice without reducing electrical conductivity. Such qualities make these materials exhibit PGEC behaviour.

The composition of skutterudites corresponds to the chemical formula LM_4X_{12}, where L is a rare earth metal, M a transition metal and X a metalloid, a group V element or pnictogen whose properties lie between those of a metal and nonmetal such as phosphorus, antimony, or arsenic. These materials could be potential in multistage thermoelectric devices as it has been shown that they have ZT>1.0, but their properties are not well known and optimization of their structures is under way.

Oxide Thermoelectrics

Because of their layered superlattice structure, homologous oxide compounds (such as those of the form $(SrTiO3)_n$(SrO)m—the Ruddleson-

Popper phase) have the potential to be used in high-temperature thermoelectric devices. These materials exhibit low thermal conductivity perpendicular to the layers while maintaining electrical conductivity within the layers. The figure of merit in oxides is still relatively low (~0.34 at 1,000K), but their enhanced thermal stability, as compared to conventional high-ZT bismuth compounds, makes them superior for use in high-temperature applications.

Interest in oxides as thermoelectric materials was reawakened in 1997 when Na_xCoO_2 was found to exhibit good thermoelectric behaviour. In addition to their thermal stability, other advantages of oxides are their nontoxicity and high oxidation resistance. Research on thermoelectric oxide materials is ongoing, but it seems that in order to simultaneously control both the electric and phonon systems, nanostructured materials are required. Some layered oxide materials are thought to have ZT~2.7 at 900 K. If the layers in a given material have the same stoichiometry, they will be stacked so that the same atoms will not be positioned on top of each other, impeding phonon conductivity perpendicular to the layers.

Half Heusler Alloys

Half Heusler alloys have potential for high temperature power generation applications especially as n-type material. These alloys have three components that originate from different element groups or might even be a combination of elements in the group. Two of the groups are composed of transition metals and the third group consists of metals and metalloids. Currently only n-type material is usable in thermoelectrics but some sources claim that they have achieved ZT~1.5 at 700 K, but according to other source only ZT~0.5 at 700 K has been achieved. They state that primary reason for this difference is the disagreement between thermal conductivities measured by different groups. These alloys are relatively cheap and also have a high power factor.

Electrically Conducting Organic Materials

Some electrically conducting organic materials may have a higher figure of merit than existing inorganic materials. Seebeck coefficient can be even millivolts per Kelvin but electrical conductivity is usually very low resulting small figure of merit. Quasi-one-dimensional organic crystals are formed from linear chains or stacks of molecules that are packed into a 3D crystal. It has theoretically been shown that under certain conditions some Q1D organic crystals may have ZT~20 at room temperature for both p- and n-type materials. In the Thermoelectrics

Handbook chapter 36.4 this has been accredited to an unspecified interference between two main electron-phonon interactions leading to the formation of narrow strip of states in the conduction band with a significantly reduced scattering rate as the mechanism compensate each other causing high ZT.

Silicon-Germanium

Silicon-germanium alloys are currently the best thermoelectric materials around 1000 ! and are therefore used in some radioisotope thermoelectric generators (RTG) (notably the MHW-RTG and GPHS-RTG) and some other high temperature applications, such as waste heat recovery. Usability of silicon-germanium alloys is limited by their high price and in addition, ZT is also only in the mid-range (~0.7).

Sodium-Cobaltate

Experiments on crystals of sodium cobaltate, using X-ray and neutron scattering experiments carried out at the European Synchrotron Radiation Facility (ESRF) and the Institut Laue-Langevin (ILL) in Grenoble were able to suppress thermal conductivity by a factor of six compared to vacancy-free sodium cobaltate. The experiments agreed with corresponding density functional calculations. The technique involved large anharmonic displacements of Na0.8CoO2 contained within the crystals.

Functionally Graded Materials

With functionally graded materials, it is possible to improve the conversion efficiency of existing thermoelectric materials. These materials have a non-uniform carrier concentration distribution and in some cases also solid solution composition. In power generation applications the temperature difference can be several hundred degrees and therefore devices made from homogeneous materials have some part that operates at the temperature where ZT is substantially lower than its maximum value. This problem can be solved by using materials whose transport properties vary along their length thus enabling substantial improvements to the operating efficiency over large temperature differences. This is possible with functionally graded materials as they have a variable carrier concentration along the length of the material, which is optimized for operations over specific temperature range.

Nanomaterials

In addition to the nanostructured Bi2Te3/Bi2Se3 superlattice thin films that have shown a great deal of promise, other nanomaterials show potential in improving thermoelectric materials. One example involving PbTe/PbSeTe quantum dot superlattices provides an enhanced ZT (approximately 1.5 at room temperature) that was higher than the bulk ZT value for either PbTe or PbSeTe (approximately 0.5). Not all nanocrystalline materials are stable, because the crystal size can grow at high temperatures ruining materials desired characteristics. In nanocrystalline material, there are many interfaces between crystals, which scatter phonons so the thermal conductivity is reduced. Phonons are confined to the grain, if their mean free path is larger than the material grain size. Measured lattice thermal conductivity in nanowires is known to depend on roughness, the method of synthesis and properties of the source material.

Nanocrystalline transition metal silicides are a promising material group for thermoelectric applications, because they fulfill several criteria that are demanded from the commercial applications point of view. In some nanocrystalline transition metal silicides the power factor is higher than in the corresponding polycrystalline material but the lack of reliable data on thermal conductivity prevents the evaluation of their thermoelectric efficiency.

One advantage of nanostructured skutterudites over normal skutterudites is their reduced thermal conductivity but further performance improvements can be achieved by using composites and by controlling the grain size, the compaction conditions of polycrystalline samples and the carrier concentration. Thermal conductivity reduction is caused by grain boundary scattering. ZT values of ~ 0.65 and >0.4 have been achieved with $CoSb_3$ based samples, the former value is 2.0 for Ni and 0.75 for Te doped material at 680 K and latter for Au-composite at T>700 K.

Due to the unique nature of graphene, engineering of thermoelectric device with extremely high Seebeck coefficient based on this material is possible. One theoretical study suggests that the Seebeck coefficient might achieve a value of 30 mV/K at room temperature and ZT for their proposed device would be approximately 20.

Superlattices and quantum wells can be good thermoelectric materials, but their production is too difficult and expensive for general use because of their fabrication is based on various thin film growth methods. Superlattice structures allow the independent manipulation of transport parameters by adjusting the structural parameters enabling the search for better understanding of thermoelectric phenomena in nanoscale. Many strategies exist to decrease the superlattice thermal conductivity that are based on engineering of phonon transport. The thermal conductivity along the film plane and wire axis can be reduced by creating diffuse interface scattering and by reducing the interface separation distance, both which are caused by interface roughness. The interface roughness can be natural due to the mixing of atoms at the interfaces or artificial. Many different structure types, such as quantum dot interfaces and thin films on step-covered substrates, can act as source for artificial roughness.

However while engineering interface structures for reduced phonon thermal conductivity effects to electron transport has to be taken into account because the reduced electrical conductivity could negate the advantage received from phonon transport engineering. Because electrons and phonons have different wavelengths, it may be possible to engineer the structure in such a way that phonons are scattered more diffusely at the interface than electrons. This would reduce the decrease of the electrical conductivity.

Second approach is to increase phonon reflectivity and therefore decrease the thermal conductivity perpendicular to interfaces. This can be achieved by increasing the mismatch between the materials. Some of these properties are density, group velocity, specific heat, and the phonon spectrum between adjacent layers. Interface roughness causes diffuse phonon scattering, which either increases or decreases the phonon reflectivity at the interfaces. Mismatch between bulk dispersion relations confines phonons and the confinement becomes more favourable as the difference in dispersion increases. The amount of confinement is currently unknown as only some models and experimental data exist. As with a previous method, the effects on the electrical conductivity have to be considered.

In order to further reduce the thermal conductivity, the localization of long wavelength phonons can be attempted with aperiodic superlattices or composite superlattices with different periodicities. In addition, defects, especially dislocations, can be used to reduce thermal conductivity in low dimensional systems.

Thermoelectric performance improvements in superlattices originate from various sources, usually at least the lattice thermal conductivity in the cross plane direction is very low but depending on the type of superlattice, the thermoelectric coefficient may also increase because the band structure changes. Low lattice thermal conductivity in superlattices is usually due to strong interface scattering of phonons. Electronic band structure in superlattices comprises the so-called minibands, which appear due to quantum confinement effects. In superlattices, electronic band structure depends on the superlattice period so that with very short period (~1 nm) the band structure approaches the alloy limit and with long period (e" ~60 nm) minibands become so close to each other that they can be approximated with a continuum. Especially in multi quantum well structures the parasitic heat conduction could cause significant performance reduction. Fortunately, the impact of this phenomenon can be reduced by choosing the distance between the quantum wells correctly.

The Seebeck coefficient can change its sign in superlattice nanowires due to the existence of minigaps as Fermi energy varies. This indicates that superlattices can be tailored to exhibit n or p-type behaviour by using the same dopants as those that are used for corresponding bulk materials by carefully controlling Fermi energy or the dopant concentration. With nanowire arrays, it is possible to exploit semimetal-semiconductor transition due to the quantum confinement and use materials that normally would not be good thermoelectric materials in bulk form. Such elements are for example bismuth. The Seebeck effect could also be used to determine the carrier concentration and Fermi energy in nanowires.

In quantum dot thermoelectrics, unconventional or nonband transport behaviour (e.g. tunneling or hopping) is necessary to utilise their special electronic band structure in the transport direction. It is possible to achieve ZT~3 at elevated temperatures with quantum dot superlattices, but they are almost always unsuitable for mass production. Bi_2Te_3/Sb_2Te_3 superlattice as a microcooler has been reported to have ZT~2.4 at 300 K.

Nanocomposites are promising material class for bulk thermoelectric devices, but several challenges have to be overcome to make them suitable for practical applications. It is not well understood why the improved thermoelectric properties appear only in certain materials with specific fabrication processes.

SrTe nanocrystals can be embedded in a bulk PbTe matrix so that rocksalt lattices of both materials are completely aligned (endotaxy) with optimal molar concentration for SrTe only 2%. This can cause strong phonon scattering but would not affect charge transport. In such case, ZT~1.7 can be achieved at 815 K for p-type material.

Recently, high ZT values in single crystal silicon nanowires have been realised. Since silicon is an earth abundant material and its processing techniques have been well developed in industry. This breakthrough may have potentially great impact in commercial applications. By varying the nanowire size and impurity doping levels, researchers from Caltech were able to achieve an approximately 100-

fold improvement of ZT values over bulk material, in single-component system of silicon nanowires for cross-sectional areas of 10 nm×20 nm and 20 nm×20 nm over a broad temperature range, including ZT<1 at 200 K. Though the mechanism behind this phenomenon is still not perfectly clear. Independent measurements of the Seebeck coefficient, the electrical conductivity and the thermal conductivity, combined with theory, indicate that the improved efficiency originates from phonon effects, namely, phonon drags. These results are expected to apply to other classes of semiconductor nanomaterials.

Tin Selenide

In 2014 a research group discovered that tin selenide (SnSe) has a ZT of 2.6 along the b axis of the unit cell. This is the highest value known so far.

Production Methods

Production methods for these materials can be divided into powder and crystal growth based techniques. Powder based techniques offer excellent ability to control and maintain desired carrier distribution. In crystal growth techniques dopants are often mixed with melt, but diffusion from gaseous phase can also be used. In the zone melting techniques disks of different materials are stacked on top of others and then materials are mixed with each other when a travelling heater causes melting. In powder techniques, either different powders are mixed with a varying ratio before melting or they are in different layers as a stack before pressing and melting.

Batteryless Radio

A batteryless radio is a radio receiver which does not require the use of a battery to provide it with electrical power.

Originally this referred to units which could be used directly by AC mains supply (mains radio); it can also refer to units which do not require a power source at all, except for the power that they receive from radio waves.

Pyroelectric Effect

Pyroelectricity (from the Greek *pyr*, fire, and electricity) is the ability of certain materials to generate a temporary voltage when they are heated or cooled. The change in temperature modifies the positions of the atoms slightly within the crystal structure, such that the polarization of the material changes. This polarization change gives rise to a voltage across the crystal. If the temperature stays constant at its new value, the pyroelectric voltage gradually disappears due to leakage current (the leakage can be due to electrons moving through the crystal, ions moving through the air, current leaking through a voltmetre attached across the crystal, etc.).

Pyroelectricity should not be confused with thermoelectricity: In a typical demonstration of pyroelectricity, the whole crystal is changed from one temperature to another, and the result is a temporary voltage across the crystal. In a typical demonstration of thermoelectricity, one part of the device is kept at one temperature and the other part at a different temperature, and the result is a *permanent* voltage across the device as long as there is a temperature difference.

The Pyroelectric Crystal Classes and Piezoelectricity

All crystal structures can be divided into 32 crystal classes, according to the number of rotational axes and reflection planes they exhibit that leave the crystal structure unchanged. Of the thirty-two crystal classes, twenty-one are non-centrosymmetric (not having a centre of symmetry). Of these twenty-one, twenty exhibit direct piezoelectricity, the remaining one being the cubic class 432. Ten of these twenty piezoelectric classes are polar, i.e., they possess a spontaneous polarization, having a dipole in their unit cell, and exhibit pyroelectricity. If this dipole can be reversed by the application of an electric field, the material is said to be ferroelectric. Any dielectric material develops a dielectric polarization (electrostatics) when an electric field is applied, but a substance which has such a natural charge separation even in the absence of a field is called a polar material. Whether or not a material is polar is determined solely by its crystal structure. Only 10 of the 32 point groups are polar. All polar crystals are pyroelectric, so the 10 polar crystal classes are sometimes referred to as the pyroelectric classes.

Piezoelectric crystal classes: 1, 2, m, 222, mm2, 4, -4, 422, 4mm, -42m, 3, 32, 3m, 6, -6, 622, 6mm, -62m, 23, -43m

Pyroelectric: 1, 2, m, mm2, 3, 3m, 4, 4mm, 6, 6mm

The property of pyroelectricity is the measured change in net polarization (a vector) proportional to a change in temperature. The total pyroelectric coefficient measured at constant stress is the sum of the pyroelectric coefficients at constant strain (primary pyroelectric effect) and the piezoelectric contribution from thermal expansion (secondary pyroelectric effect). Under normal circumstances, even polar materials do not display a net dipole moment. As a consequence there are no electric dipole equivalents of bar magnets because the intrinsic dipole moment is neutralized by "free" electric charge that builds up on the surface by internal conduction or from the ambient atmosphere. Polar crystals only reveal their nature when perturbed in some fashion that momentarily upsets the balance with the compensating surface charge.

Recent Developments

Progress has been made in creating artificial pyroelectric materials, usually in the form of a thin film, out of gallium nitride (GaN), caesium nitrate ($CsNO_3$), polyvinyl fluorides, derivatives of phenylpyridine, and cobalt phthalocyanine. Lithium tantalate ($LiTaO_3$) is a crystal exhibiting both piezoelectric and pyroelectric properties, which has been used to create small-scale nuclear fusion ("pyroelectric fusion").

Mathematical Description

The pyroelectric coefficient may be described as the change in the spontaneous polarization vector with temperature:

$$p_i = \frac{\partial P_{S,i}}{\partial T}$$

where p_i ($Cm^{-2}K^{-1}$) is the vector for the pyroelectric coefficient.

Power Generation

A pyroelectric can be repeatedly heated and cooled (analogously to a heat engine) to generate usable electrical power. One group calculated that a pyroelectric in an Ericsson cycle could reach 50% of Carnot efficiency, while a different study found a material that could in theory reach 84-92% of Carnot efficiency (these efficiency values are for the pyroelectric itself, ignoring losses from heating and cooling the substrate, other heat-transfer losses, and all other losses elsewhere in the system). Possible advantages of pyroelectric generators for generating electricity (as compared to the conventional heat engine plus electrical generator) include potentially lower operating temperatures, less bulky equipment, and fewer moving parts. Although a few patents have been filed for such a device, it does not appear to be anywhere close to commercialization yet.

Thermionic Converter

A thermionic converter consists of a hot electrode which thermionically emits electrons over a potential energy barrier to a cooler electrode, producing a useful electric power output. Caesium vapor is used to optimize the electrode work functions and provide an ion supply (by surface ionization or electron impact ionization in a plasma) to neutralize the electron space charge.

The scientific aspects of thermionic energy conversion primarily concern the fields of surface physics and plasma physics. The electrode surface properties determine the magnitude of electron emission current and electric potential at the electrode surfaces, and the plasma properties determine the transport of electron current from the emitter to the collector. All practical thermionic converters to date employ caesium vapor between the electrodes, which determines both the surface and plasma properties. Caesium is employed because it is the most easily ionized of all stable elements.

The surface property of primary interest is the work function, which is the barrier that limits electron emission current from the surface and essentially is the heat of vaporization of electrons from the surface. The work function is determined primarily by a layer of caesium atoms adsorbed on the electrode surfaces. The properties of the interelectrode plasma are determined by the mode of operation of the thermionic converter. In the ignited (or "arc") mode the plasma is maintained via ionization internally by hot plasma electrons (~ 3300 K); in the unignited mode the plasma is maintained via injection of externally produced positive ions into a cold plasma; in the hybrid mode the plasma is maintained by ions from a hot-plasma interelectrode region transferred into a cold-plasma interelectrode region.

Recent Work

All the applications cited above have employed technology in which the basic physical understanding and performance of the thermionic converter were essentially the same as those achieved before 1970. During the period from 1973 to 1983, however, significant research on advanced low-temperature thermionic converter technology for fossil-fuelled industrial and commercial electric power production was conducted in the US, and continued until 1995 for possible space reactor and naval reactor applications. That research has shown that substantial improvements in converter performance can be obtained now at lower operating temperatures by addition of oxygen to the caesium vapor, by suppression of electron reflection at the electrode surfaces, and by hybrid mode operation. Similarly, improvements via use of oxygen-containing electrodes have been demonstrated in Russia along with design studies of systems employing the advanced thermionic converter performance. Recent studies have shown that excited Cs-atoms in thermionic converters form clusters of Cs-Rydberg matter which yield a decrease of collector emitting work function from 1.5 eV to 1.0 – 0.7 eV. Due to long-lived nature of Rydberg matter this low work function remains low for a long time which essentially increases the low-temperature converter's efficiency.

Synthesis of Size Controlled Copper Nanoparticles

Nanomaterials are being applied in more and more fields within engineering and technology. One of the key benefits of nanomaterials is that their properties differ from bulk material of the same composition. The properties of nanoparticles, for example, can be easily altered by varying their size, shape, and chemical environment.

Copper is a Block D, Period 4 element. It is a ductile metal with very high thermal and electrical conductivity. The morphology of copper nanoparticles is round, and they appear as a brown to black powder.

Copper is found to be too soft for some applications, and hence it is often combined with other metals to form numerous alloys such as brass, which is a copper-zinc alloy.

Copper nanoparticles are graded as highly flammable solids, therefore they must be stored away from sources of ignition. They are also known to be very toxic to aquatic life.

Chemical Properties

The chemical properties of copper nanoparticles are outlined in the following table.

Chemical Data	
Chemical symbol	Cu
CAS No.	7440-50-8
Group	11
Electronic configuration	[Ar] $3d^{10}$ $4s^{1}$

Physical Properties

The physical properties of copper nanoparticles are given in the following table.

Properties	Metric	Imperial
Density	8.94 g/cm^3	0.00032 lb/in^3
Molar mass	63.55 g/mol	-

Thermal Properties

The thermal properties of copper nanoparticles are provided in the table below.

Properties	Metric	Imperial
Melting point	1083°C	1981.4°F
Boiling point	2567°C	4652.6°F

Manufacturing Process

Copper nanoparticles can be manufactured using numerous methods. The electrodeposition method is considered by many as one of the most suitable and easiest. The electrolyte used for the process is an acidified aqueous solution of copper sulfate with specific additives. A spongy layer of copper particles is deposited on the cathode surface when the input DC voltage is varied with a constant current. The particles are typically characterized and assessed by XRD and UV-Vis. The surface morphological characterization is done using scanning electron microscopy (SEM) and transmission electron microscopy (TEM).

Damp reunion tends to affect the dispersion performance and usable properties of copper nanoparticles; hence this material has to be sealed under vacuum and stored in a cool and dry room. It should not be exposed to air, and should not be under stress.

Applications

The key applications of copper nanoparticles are listed below:

- Acts as an anti-biotic, anti-microbial, and anti-fungal agent when added to plastics, coatings, and textiles
- Copper diet supplements with efficient delivery characteristics
- High strength metals and alloys
- EMI shielding
- Heat sinks and highly thermal conductive materials
- Efficient catalyst for chemical reactions and for the synthesis of methanol and glycol
- As sintering additives and capacitor materials
- Conductive inks and pastes containing Cu nanoparticles can be used as a substitute for very expensive noble metals used in printed electronics, displays, and transmissive conductive thin film applications
- Superficial conductive coating processing of metal and non-ferrous metal
- Production of MLCC internal electrode and other electronic components in electronic slurry for the miniaturization of microelectronic devices;
- As nanometal lubricant additives

Copper nanoparticles application research is ongoing to discover their potential dielectric, magnetic, electrical, optical, imaging, catalytic, biomedical and bioscience properties.

Synthesis and Thermal Processing of Copper Nanoparticles

Copper nanoparticles were synthesized by thermal decomposition using copper chloride, sodium oleate, and phenyl ether as solvent agents. The formation of nanoparticles was evidenced by the X-ray diffraction and transmission electron microscopy. The peaks in the XRD pattern correspond to the standard values of the face centred cubic (fcc) structure of metallic copper and no peaks of other impurity crystalline phases were detected. TEM analysis showed spherical nanoparticles with sizes in the range of 4 to 18nm. The antibacterial properties of copper nanoparticles were evaluated in vitro against strains of Staphylococcus aureus and Pseudomonas aeruginosa. The antibacterial activity of copper nanoparticles synthesized by thermal decomposition showed significant inhibitory effect against these highly multidrug-resistant bacterial strains.

Recently metallic nanoparticles (NPs) have attracted great interest because of their unique physical and chemical properties. Their properties can be controlled depending on the synthesis method. One of the main effects, which are enhanced by controlling particle size, is their antimicrobial action. The antimicrobial activity of NPs is known to be a function of surface area in contact with microorganisms. For this reason, high surface area NPs assure a wide range of reactions on the surface of microorganisms, inhibiting the normal function of cells or causing cell death.

Several methods have been used for the formation of copper nanoparticles. One of the first reported methods to obtain inorganic NPs is the chemical coprecipitation method, which involves nucleation and growth during the same process [4], but the main disadvantage of this process is the formation of agglomerated NPs with a wide size distribution. Nanometric particles with size control have been obtained through homogeneous precipitation reactions, which involve separation of nucleation and growth process. Under conditions of homogeneous precipitation, an abrupt nucleation occurs when reactive concentrations reach a supercritical saturation point [5], generating this way a homogeneous growth by solutes diffusion from solution to surface, achieving their final size. The key to obtain homogeneous particle size is the separation of nucleation and growth steps, avoiding nucleation during NPs growth step; if performed with direct heating of the mixture, a wide size distribution of NPs can be produced.

Thermal decomposition of metallic precursors in presence of organic surfactants at high temperature is a method widely used due to its ease to produce highly crystalline NPs with size distribution control. In the particular case of Cu NPs, low redox potential causes surface oxidation which decreases their antimicrobial characteristics, this is due to the fact that metallic copper (Cu0) is more reactive than copper oxide. Therefore, in order to obtain Cu NPs with high antimicrobial properties, the size control of NPs and the presence of a surface coating to prevent their oxidation are very important.

Synergistic Activity of GoldPlatinum Alloy Nanoparticle Catalysts

Platinum is a chemical element with the chemical symbol Pt and an atomic number of 78. Gold is popular because it can be worked into almost any shape. Yellow gold jewellery of 18K and above does not tarnish and rarely causes problems for people with skin irritations. White Gold is popular for its appearance and price point compared to platinum alloys.

Technically there is no such thing as 'White Gold.' Gold can be lightened by combining it with light metals but we plate all WG pieces with Rhodium; a member of the platinum family and the whitest precious metal after silver. This rhodium plating creates a hard skin with good resistance. Over time plating may wear through. Re-plating is a fairly simple process, depending on the condition of the piece. In most cases this will be done approximately as often as a platinum ring requires re-polishing, although a fine plating job may last longer than a polish on platinum due to the superior hardness of rhodium.

Platinum is a versatile, eternal metal. It does not fade or tarnish and is ideal for those with sensitive skin because it is hypoallergenic. It is the safest, most suitable and versatile metal for durably setting any kind of gemstone. Platinum's density gives it a unique quality. When platinum scratches none of the volume is lost, the metal is merely displaced as ridges are raised on the edge of the scratch. As platinum is worn it develops a patina-like appearance. It can be polished again and again because this is just moving the metal around, not wearing it down. Other precious metals lose material over time. Gold prongs wear down and rings can get thinner with wear. Platinum prongs bend but rarely break and do not wear down.

Common Platinum Alloys

There are 4 platinum alloys commonly used in the USA.

1. Pt900/Ir = 900 parts platinum, 100 parts Iridium
2. Pt950/Ir = 950 parts platinum, 50 parts Iridium
3. Pt950/Ru = 950 parts platinum, 50 parts Ruthenium
4. Pt950/Co = 950 parts platinum, 50 parts Cobalt

A Common Misunderstanding

When people hear Pt950 described as 95% platinum they assume that means 95% by volume. It doesn't. The percentage is by weight. Platinum is the heaviest of these metals. Therefore, it will require more than 5% *by volume* of a lighter metal to match platinum by 5% *in weight.* This misunderstanding is merely academic information that may be of interest to some people.

Platinum and to a much greater extent palladium are the principal platinum group metals used in dental restorations. The metals are usually mixed with gold or silver as well as copper and zinc in varying ratios to produce alloys suitable for dental inlays, crowns and bridges. Small amounts of ruthenium or iridium are sometimes added.

The most common application is in crowns, where the alloy forms the core onto which porcelain is bonded to build up an artificial tooth. The aim of using platinum group metals in dental alloys is to provide strength, stiffness and durability whilst the other alloyed metals provide malleability.

Alloy Types

There are two main types of precious metal alloys used in dentistry:

- High gold alloys, normally alloyed with a small amount (around 10%) of platinum
- Low gold alloys, predominantly palladium based with a palladium content typically ranging from 50% to 80%

Development of the Market for Palladium

Platinum-containing high gold alloys have been used by dentists for many decades but the use of palladium in dentistry is relatively recent. It dates from the 1980s, when a rise in the price of gold encouraged palladium to be introduced as a lower-cost alternative.

In Japan, the government operates a specific mandate stating that all government-subsidised dental alloys have to include a palladium content of 20%. This alloy is known as the kinpala alloy and is used in around 90% of all Japanese dental treatment. Hence, Japan is the largest palladium-consuming region for dental applications, followed by North America and then Europe.

Core-Shell Nanoparticles as Electrocatalysts

Core–shell semiconducting nanocrystals (CSSNCs) are a class of materials which have properties intermediate between those of small, individual molecules and those of bulk, crystalline semiconductors.

They are unique because of their easily modular properties, which are a result of their size. These nanocrystals are composed of a quantum dot semiconducting core material and a shell of a distinct semiconducting material.

The core and the shell are typically composed of type II–VI, IV–VI, and III–V semiconductors, with configurations such as CdS/ZnS, CdSe/ZnS, CdSe/CdS, and InAs/CdSe (typical notation is: core/shell) Organically passivated quantum dots have low fluorescence quantum yield due to surface related trap states. CSSNCs address this problem because the shell increases quantum yield by passivating the surface trap states.

In addition, the shell provides protection against environmental changes, photo-oxidative degradation, and provides another route for modularity. Precise control of the size, shape, and composition of the core and shell enable the emission wavelength to be tuned over a wider range of wavelengths than with either individual semiconductor. These materials have found applications in biological systems and optics.

Colloidal semiconductor nanocrystals, which are also called quantum dots(QDs), consist of ~1–10 nm diameter semiconductor nanoparticles that have organic ligands bound to their surface. These nanomaterials have found applications in nanoscale photonic, photovoltaic, and light-emitting diode (LED) devices due to their size-dependent optical and electronic properties. Quantum dots are popular alternatives to organic dyes as fluorescent labels for biological imaging and sensing due to their small size, tuneable emission, and photostability.

The luminescent properties of quantum dots arise from exciton decay (recombination of electron hole pairs) which can proceed through a radiative or nonradiative pathway. The radiative pathway involves electrons relaxing from the conduction band to the valence band by emitting photons with wavelengths corresponding to the semiconductor's bandgap.

Nonradiative recombination can occur through energy release via phonon emission or auger recombination. In this size regime, quantum confinement effects lead to a size dependent increasing bandgap with observable, quantized energy levels.

The quantized energy levels observed in quantum dots lead to electronic structures that are intermediate between single molecules which have a single HOMO-LUMO gap and bulk semiconductors which have continuous energy levels within bands.

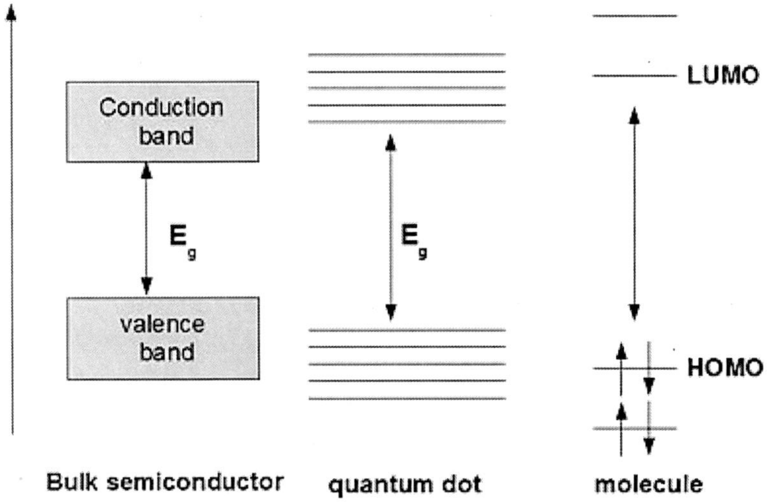

Semiconductor nanocrystals generally adopt the same crystal structure as their extended solids. At the surface of the crystal, the periodicity abruptly stops, resulting in surface atoms having a lower coordination number than the interior atoms. This incomplete bonding (relative to the interior crystal structure) results in atomic orbitals that point away from the surface called "dangling orbitals" or unpassivated orbitals. Surface dangling orbitals are localized and carry a slight negative or positive charge.

Weak interaction among the inhomogeneous charged energy states on the surface has been hypothesized to form a band structure. If the energy of the dangling orbital band is within the semiconductor bandgap, electrons and holes can be trapped at the crystal surface. For example, in CdSe quantum dots, Cd dangling orbitals act as electron traps while Se dangling orbitals act as hole traps. Also, surface defects in the crystal structure can act as charge carrier traps.

Charge carrier trapping on QDs increases the probability of non-radiative recombination, which reduces the fluorescence quantum yield. Surface-bound organic ligands are typically used to coordinate to surface atoms having reduced coordination number in order to passivate the surface traps.

For example, tri-n-octylphosphine oxide (TOPO) and trioctylphospine(TOP) have been used to control the growth conditions and passivate the surface traps of high quality CdSe quantum dots. Although this method provides narrow size distributions and good crystallinity, the quantum yields are ~5–15%. Alkylamines have been incorporated into the TOP/TOPO synthetic method to increase the quantum yields to ~50%.

The main challenge in using organic ligands for quantum dot surface trap passivation is the difficulty in simultaneously passivating both anionic and cationic surface traps. Steric hindrance between bulky organic ligands results in incomplete surface coverage and unpassivated dangling orbitals.

Growing epitaxial inorganic semiconductor shells over quantum dots inhibits photo-oxidation and enables passivation of both anionic and cationic surface trap states. As photogenerated charge carriers are less likely to be trapped, the probability for excitons to decay through the radiative pathway increases. CdSe/CdS and ZnSe/CdSe nanocrystals have been synthesized that exhibit 85% and 80–90% quantum yield, respectively.

Core–shell semiconductor nanocrystal architecture was initially investigated in the 1980s, followed by a surge of publications on synthetic methods the 1990s.

Classification of Core–Shell Semiconductor Nanocrystals

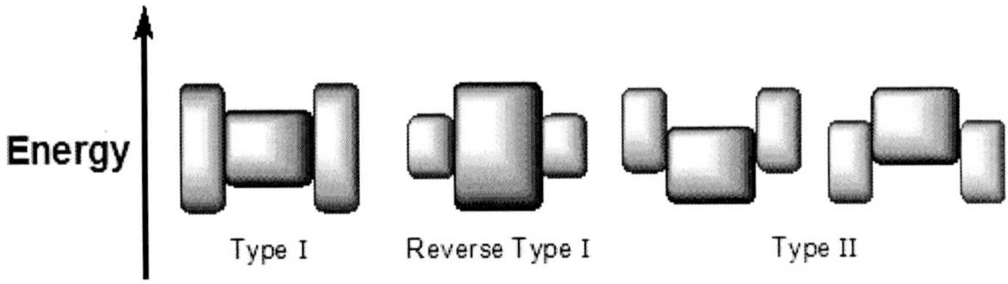

Figure: *The three types of core–shell nanocrystals. The upper and lower edges represent the upper and lower energy edges of the core (blue) and the shell (red).*

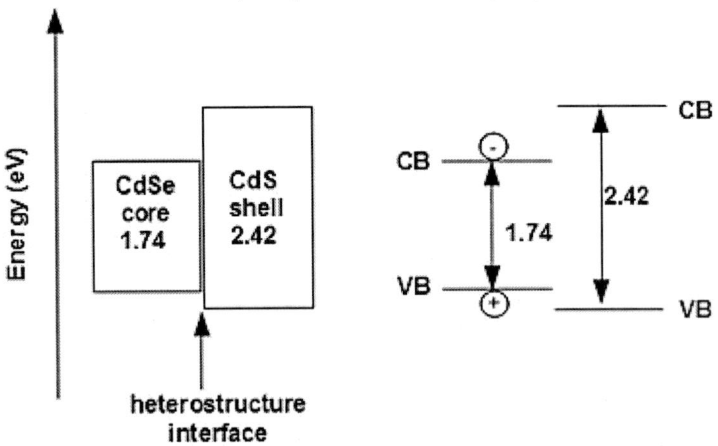

Figure: *Type I core shell semiconductor nanocrystal (CdSe/CdS) band-edge alignment. VB=valence band, CB=conduction band*

Core shell semiconductor nanocrystal properties are based on the relative conduction and valence band edge alignment of the core and the shell. In type I semiconductor heterostructures, the electron and holes tend to localize within the core. In type II heterostructures, one carrier is localized in the shell while the other is localized in the core.

Type I

Description

In a Type I CSSNC, the bandgap of the core is smaller than that of the shell. Both the conduction and valence band edges of the core lie within the bandgap of the shell, which confines both electrons and holes in the core. This can be seen in figure X, where the electron and hole of an exciton at the CdSe(bandgap:1.74 eV)/CdS(bandgap:2.42 eV) interface occupy energy states within the CdSe core, which corresponds to the lowest available energy separation. The emission wavelength due to radiative electron-hole recombination within the core is slightly redshifted compared to uncoated CdSe.

- Examples

CdSe/CdS, CdSe/ZnS, and InAs/CdSe

Reverse Type I

- Description

In the reverse type I configuration, the core has a wider bandgap than the shell, and the conduction and valence band edges of the shell lie within those of the core. The lowest available exciton energy separation occurs when the charge carriers are localized in the shell. Changing the shell thickness tunes the emission wavelength.

- Examples

CdS/HgS, CdS/CdSe, and ZnSe/CdSe

Type II

- Description

In the type II configuration, the valence and conduction band edge of the core are both lower or higher than the band edges of the shell. An example of a type II is shown in figure X, ZnTe(bandgap:2.26)/CdSe(bandgap:1.74). The lowest energy separation of the electron and the hole will occur when the hole is confined in the ZnTe core valence band and the electron is confined in the CdSe shell conduction band. The emission wavelength will be determined by the energy difference between these occupied states, as shown by the red arrow, which will be at a lower energy than either of the individual bandgaps. The emission wavelength can be significantly red shifted compared to the unpassivated core.

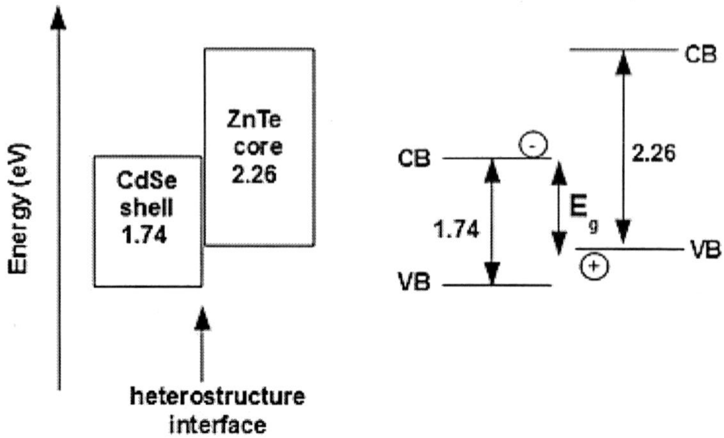

- Examples

ZnTe/CdSe, CdTe/CdSe, CdS/ZnSe

Doped Core Shell Semiconductor Nanocrystals

Doping has been shown to strongly affect the optical properties of semiconductor nanocrystals. Impurity concentrations in semiconductor nanocrystals grown using colloidal synthesis, however, are typically lower than in their bulk counterparts. There has been interest in magnetic doping of CSSNCs for applications in magnetic memory and spin-based electronics. Dual-mode optical and magnetic resonance (MR) imaging has been explored by doping the shell of CdSe/ZnS with Mn, which caused the CSSNC to be paramagnetic.

Synthesis

In synthesizing core shell nanoparticles, scientists have studied and found several wet chemical methods, such as chemical precipitation, sol-gel, microemulsion and inverse micelle formation. Those methods have been used to grow core shell chalcogenide nanoparticles with an emphasis on better control of size, shape, and size distribution. To control the growth of nanoparticles with tunable optical properties, supporting matrices such as glasses, zeolites, polymers or fatty acids have been used. In addition, to prepare nanoparticles of sulfides, selenides and tellurides, the Langmuir–Blodgett film technique has been used successfully. In comparison to wet chemical methods, electrochemical synthesis is more desirable, such as the use of aqueous solvents rather than toxic organic solvents, formation of conformal deposits, room-temperature deposition, low cost, and precise control of composition and thickness of semiconductor coating on metal nanoparticles. However, owing to the difficulty of preparing electrically addressable arrays of nanoparticles, the use of electrochemical techniques to produce core-shell nanoparticles was difficult. Recently, Cadmium Sulfide (CdS) and Copper iodide (CuI) was electrochemically grown on a 3-D nanoelectrode array via layer-by-layer depositing of alternating layers of nanoparticles and Polyoxometalate(POM).

CSSNCs can be grown by using collodial chemistry methods with an appropriate control of the reaction kinetics. Using this method which results in a relatively high control of size and shape, semiconductor nanostructures could be synthesized in the form of dots, tubes, wires and other forms which show interesting optic and electronic size-dependent properties.

Since the synergistic properties resulting from the intimate contact and interaction between the core and shell, CSSNCs can provide novel functions and enhanced properites which are not observed in single nanoparticles.

The size of core materials and the thickness of shell can be controlled during synthesis. For example, in the synthesis of CdSe core nanocrystals, the volume of H_2S gas can determine the size of core nanocrystals. As the volume of H_2S increases, the size of the core decreases. Alternatively, when the reaction solution reaches the desired reaction temperature, rapid cooling can result in smaller core sizes. In addition, the thickness of shell is typically determined by the added amount of shell material during the coating process.

Characterization

An increase in either the core size or shell length results in longer emission wavelengths. The interface between the core and shell can be tailored to passivate relaxation pathways and form radiative states. The size dependence of the band gap in these nanoparticles due to the quantum confinement effect has been utilised to control the photoluminescence colour from blue to red by preparing nanoparticles of varying sizes. By manipulating the size or shape of the nanoparticles, the luminescence colours and purity can be controlled. However, the quantum yield and the brightness of luminescence of the CSSNCs is ultimately limited and it cannot be controlled because of the presence of surface traps.

UV-vis absorption spectra, X-ray diffraction (XRD), transmission electron microscopy (TEM) and X-ray photoelectron spectroscopy (XPS) are the techniques typically used to identify and characterize CSSNCs.

Applications

One of the most important properties of core–shell semiconducting nanocrystals (CSSNCs) is that their cores, which are quantum dots, fluoresce, which is important in their biomedical and optical applications. The shells are highly modular, and thus the bulk properties, such as solubility and activity of the CSSNCs can be changed.

Biomedical Applications

The properties desired of CSSNCs when using them for biological applications include high quantum yield, narrow fluorescence emission, broad absorption profile, stability against photobleaching, 20 second fluorescent lifetime, and high brightness. High quantum yields mean that minimal energy will need to be put into the quantum dot to induce fluorescence. A narrow fluorescence emission allows for multiple colours to be imaged at once without colour overlap between different types of CSSNCs. Having a broad absorption profile allows multiple CSSNCs to be excited at the same wavelength and thus, multiple CSSNCs could be imaged simultaneously.

Having a 20 second fluorescent lifetime allows for time-resolved bioimaging. The utility of CSSNCs is that they can be a complement to organic fluorophores. CSSNCs are less susceptible to photobleaching, but less is known about them compared to organic fluorophores. CSSNCs have 100–1000 times the two-photon fluorescence efficiency as organic dyes, exemplifying their value. In the cases where CSSNCs are used in biological medium, the core is a quantum dot and the shell can be an organic molecule or biological ligands, such as a DNA, that are used for biocompatibility and targeting. The shell can also be an organic molecule to which a biological molecule is later conjugated, furthering the modularity of core–shell structure. The most popular core/shell pair used is CdSe core with ZnS or CdS shell, which improves the quantum yield and protects against photobleaching compared to that of the core material alone. The size of the CSSNC is directly correlated to the colour of fluorescence, so being able to control particle size is desirable. However, it is generally unknown how the shell molecules, and salt concentration, pH, and temperature of the media affect the CSSNCs' properties and remains empirical.

In Vitro Cell Labelling

Because multiple colours can be imaged, CSSNCs' ability to be used in cell labelling is of growing importance. However, it can be difficult to get CSSNCs across the cell membrane. This has been achieved via endocytosis (the most common method), direct microinjection, and electroporation, and once in the cell, they become concentrated in the nucleus and can stay there for extended periods of time.

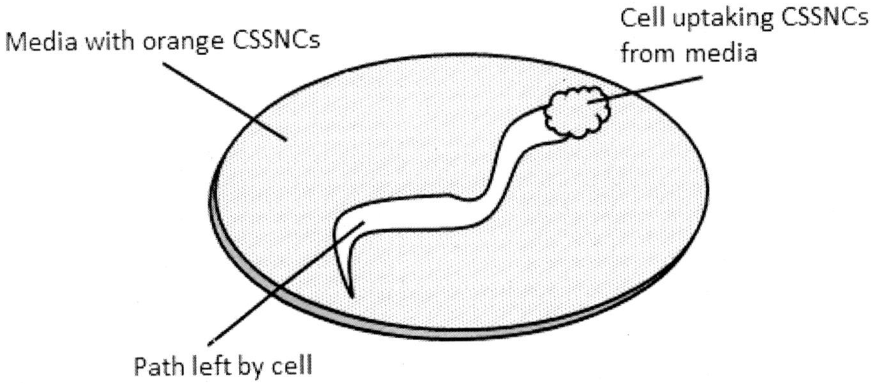

Once CSSNCs are inside cells, they remain even after cellular division and can be imaged in both mother and daughter cells. This particular technique was shown using *Xenopus* embryos. Another example of CSSNCs is seen in their tracking ability; when cells are gown on a 2D matrix embedded with CSSNCs, cells uptake the CSSNCs as they move, leaving a trail seen as the absence of CSSNCs. This means that the mobility of cells can be imaged, which is important since the metastatic potential of breast tissue cells has been shown to increase with mobility. Also, it has been shown that five different toxins can be detected using five different CSSNCs simultaneously. In a move toward environmentally friendlier and less toxic CSSNCs, Si quantum dots with various shells have been developed. Si is 10 times safer than Cd and current work is focused on making Si more water soluble and biocompatible. In particular, Si quantum dots with poly(acrylic acid) and allylamine shells have been used in cell labelling. Other in vitro uses include flow cyclometry, pathogen detection, and genomic and proteomic detection.

In Vivo and Deep Tissue Imaging

Because CSSNCs emit in the near-infrared region (700–900 nm) of the electromagnetic spectrum, imaging them is not complicated by autofluorescence of tissue, which occurs at higher frequencies (400–600 nm), and scattering effects. This has been used in the mapping of sentinel lymph-nodes in cancer surgery in animals. Lymph nodes 1 cm deep were imaged and the excised nodes with CSSNC accumulation were found to have the highest probability for containing metastatic cells. In addition, CSSNCs have been shown to remain fluorescent in cells in vivo for 4 months. To track and diagnose cancer cells, labelled squamous carminoma cell-line U14 cells were used and fluorescent images could be seen after 6h. CSSNCs conjugated to doxorubicin were also used to target, image, and sense prostate cancer cells that express the prostate-specific membrane antigen protein. Using a cancer-specific antibody conjugated to QDs with polymer shells is the most popular in tumor targeted imaging. The main disadvantage of using CSSNCs for in vivo imaging is the lack of information about their excretion and toxicity. The typical cores used show DNA damage and toxicity toward liver cells, but using shells seems to diminish this effect. The use of other substances in the core, such as rare-earth elements and Si, are being explored to reduce toxicity. Other disadvantages include limited commercial availability, variability in surface chemistry, nonspecific binding, and instrument limitation.

Optics

The size, shape, and composition of the core–shell structure are related to the bandgap, which in turn is related to its optical properties. Thus, by modulating the size, shape, and material of the core, the optics can be tuned and optimized for use in optical devices and applications such as LEDs, detectors, lasers, phosphors, and photovoltaics.

LEDs

Currently, CSSNC LED efficiency is less than that of organic LEDs. However, studies show that they have potential to accomplish what organic LEDs cannot. CSSNC LEDs constructed using multiple layers of CSSNCs resulted in poor conduction, charge imbalance, low luminescence efficiency, and a large number of pinhole defects. LEDs constructed of one monolayer avoid these problems. An advantage of CSSNC LEDs over organic LEDs is that CSSNC LEDs have narrower emissions, as narrow as 32 nm, than organic LEDs, which range from 50–100 nm.

Specifically, the core–shell motif is desirable for use in LEDs because of their electroluminescence and photoluminescence quantum efficiencies and their ability to be processed into devices easily. Current aims for LED displays include developing materials with wavelength emissions of 610–620 nm for red displays, 525–530 nm for green displays, and 460–470 nm for blue displays. This is because these wavelengths maximize the perceived power and they lie outside of the National Television System Committee standard colour triangle. CSSNCs have been synthesized that meet these wavelength emissions: (CdSe)ZnS for red emission, (CdS)ZnS for blue emission, and $(Cd_xZn_{1-x}Se)Cd_yZn_{1-y}S$ for the green emission. Using CdSe core and ZnS or CdS/ZnS shells, the maximum luminance values of red, orange, yellow and green LEDs were improved to 9,064, 3,200, 4,470 and 3,700 cd m"2, respectively; electroluminescent efficiency (1.1–2.8 cd A21), and turn-on voltages (3–4 V) were also increased.

Lasers

In CSSNCs with only one exciton, absorption and stimulated emission occur equally and in CSSNCs with more than one exciton, non-radiative Auger recombination occurs, which decays optical gain, an important quality in lasers. However, type II CSSNCs, CdS/ZnSe, were used in optical amplification from stimulated emission of single-exiton states, eliminating Auger recombination. This has the advantage that lasing threshold could be lowered under continuous wave excitation, enhancing the potential of CSSNCs as optical gain media. Type II CSSNCs separate the electrons and holes of the exciton pair, which leads to a strong electric field and thus, reducing absorption losses.

Phosphors

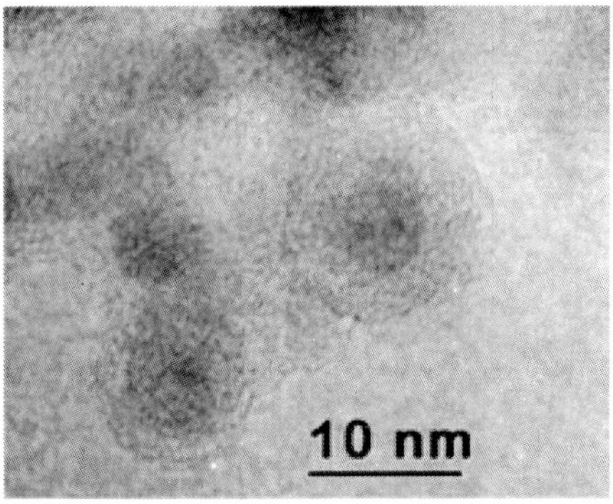

Figure: *TEM-micrograph of polymer coated g-Fe_2O_3 nanoparticles. The ceramic core appear dark, the polymer shell is visible as a bright seam around the Fe_2O_3-cores.*

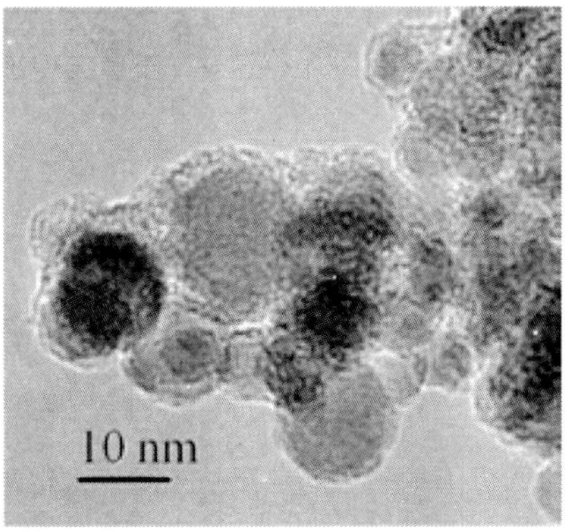

Figure: *TEM-micrograph showing ZrO_2/Fe_2O_3 core/shell nanoparticles. The ZrO_2-cores appear darker than the Fe_2O_3-shell.*

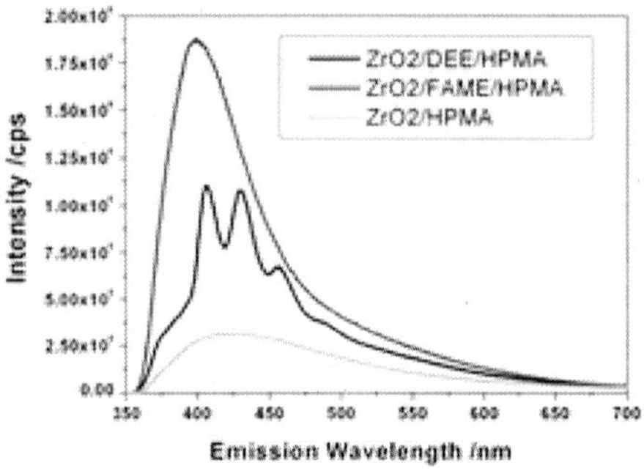

Figure: *UV-fluorescence spectra of ZrO_2 nanoparticles, coated with different organic molecules, respectively. Intensity as well as feature of the spectra depend on the type of organic molecule used. A significant difference in the spectra is visible (DEE= diethylether, FAME= formic acid methylester, HPMA= hydroxy poly methylmethacrylate). The change of optical properties is a direct proof for successful coating.*

By combining the modularity of CSSNCs and stability of organic polymer, a broad range of colours of phosphors were developed. CdSe core/ZnS shell CSSNCs are used to generate bluish green to red colours, and (CdS)ZnS QDs are used to generate violet to blue colours. By mixing the appropriate amounts of the different sizes of CSSNCs, the entire visible range with narrow emission profiles and high photoluminescence quantum yields can be achieved.Nanocomposite particles are, to our understanding, core/shell nanoparticles. They comprise a core made from one material, and a shell (or coating) made from second material. The core/shell nanoparticles we talk about are always made from an inorganic core (i.e. oxide, nitride, carbide). The shell is made from another oxide, nitride, carbide, or an organic material (monomer, surfactant, surface active organic molecule, orgnic dye). By clever selection of core and shell materials properties can be combined, or the surface can be functionalized. The particles are synthesized by the Karlsruhe Microwave Plasma Process (KMPP).

The most important and complementary tools for the characterization of core/shell nanoparticles are transmission electron microscopy (TEM) and X-ray photon sectroscopy (XPS).

CO Adsorption on Supported Metal Nanoparticle Catalysts

Of all of the reactions examined on supported metal nanoparticle catalysts, the oxidation of CO has received extensive experimental and theoretical attention.

While the reaction may appear to be rather straightforward, an understanding of the mechanism has been surprisingly difficult to achieve. Not only is the oxidation of CO a model system to study, but the adsorption of CO on gold, a fundamental step in the mechanism, has also become a model interaction to investigate.

Multiple studies indicate that CO is only weakly chemisorbed on gold. The adsorption is believed to involve low-coordinate sites with coordination number CN = 6 (i.e., corners or kinks) and CN = 7 (i.e., edges or steps).

Adsorption to terrace sites with CN = 8 or 9 does not occur.

CO adsorption studies also display a number of coverage-dependent results. For example, quantification of the coverage with CO pressure (or with temperature) reveals coverage-dependent, non-Langmuir behaviour.

The adsorption enthalpy shows a coverage-dependence, decreasing with increasing coverage.

Also, the infrared studies reveal a common CO peak that typically red shifts with increasing coverage.

To account for the physicochemical, coverage-dependent behaviour of CO adsorption on gold, we developed a treatment of the Temkin adsorption model.

This thermodynamic model is an extension of the Langmuir adsorption model that incorporates a linear variation in binding energy with coverage, and has three cases that are similar but not equivalent. The adsorbate interaction case takes into account direct adsorbate "adsorbate interactions or indirect adsorbate "substrate interactions. This case assumes that these interactions produce alinear variation of adsorption enthalpy with adsorbate coverage.

In contrast, the heterogeneous surface case assumes a uniform distribution of heterogeneous binding sites. It is assumed that the adsorption enthalpy varies linearly over these different binding sites. The third case involves a common approximation

for midrange adsorbate coverage. According to this approximation, the expressions for the adsorbate interaction and the heterogeneous surface cases are simplified, producing a newcommon expression. This common expression is the familiar Temkin isothermal result; coverage varies with the logarithm of pressure.

Magic Number

In nuclear physics, a magic number is a number of nucleons (either protons or neutrons) such that they are arranged into complete shells within the atomic nucleus. The seven most widely recognised magic numbers as of 2007 are 2, 8, 20, 28, 50, 82, and 126 (sequence A018226 in OEIS). Atomic nuclei consisting of such a magic number of nucleons have a higher average binding energy per nucleon than one would expect based upon predictions such as the semi-empirical mass formula and are hence more stable against nuclear decay.

The unusual stability of isotopes having magic numbers means that transuranium elements can be created with extremely large nuclei and yet not be subject to the extremely rapid radioactive decay normally associated with high atomic numbers (as of 2007, the longest-lived known isotope among all of the elements between 110 and 118 lasts only 8.9 min and the next 3.7 min). Large isotopes with magic numbers of nucleons are said to exist in an island of stability. Unlike the magic numbers 2–126, which are realised in spherical nuclei, theoretical calculations predict that nuclei in the island of stability are deformed. Before this was realised, higher magic numbers, such as 184 and 258, were predicted based on simple calculations that assumed spherical shapes. It is now believed that the sequence of spherical magic numbers cannot be extended in this way.

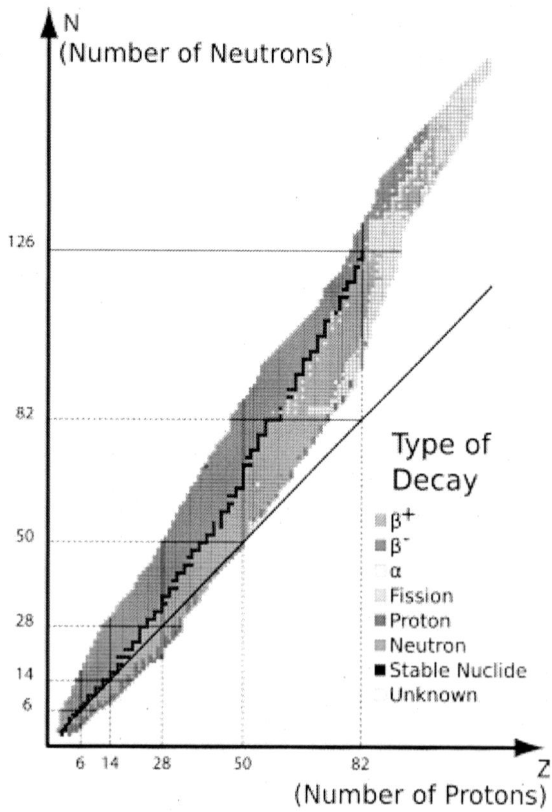

Figure: *Graph of isotope stability.*

According to Steven A. Moszkowski (a student of Maria Goeppert-Mayer), the term "magic number" was coined by Eugene Wigner: "Wigner, too, believed in the liquid drop model, but he recognised, from the work of Maria Mayer, the very strong evidence for the closed shells. It seemed a little like magic to him, and that is how the words 'Magic Numbers' were coined."

Double Magic

Nuclei which have neutron number and proton (atomic) numbers each equal to one of the magic numbers are called "double magic", and are especially stable against decay. Examples of double magic isotopes include helium-4 (4He), oxygen-16 (16O), calcium-40 (40Ca), nickel-48 (48Ni), nickel-56(^{56}Ni) and lead-208(208Pb).

Double-magic effects may allow existence of stable isotopes which otherwise would not have been expected. An example is calcium-40 (40 Ca), with 20 neutrons and 20 protons, which is the heaviest stable isotope made of the same number of protons and neutrons. Both calcium-48 (48Ca) and nickel-48 (48Ni) are double magic because calcium-48 has 20 protons and 28 neutrons while nickel-48 has 28 protons and 20 neutrons. Calcium-48 is very neutron-rich for such a light element, but like calcium-40, it is made stable by being double magic. Similarly, nickel-48, discovered in 1999, is the most proton-rich isotope known beyond helium-3.

Magic number shell effects are seen in ordinary abundances of elements: it is no accident that helium-4 (4He) is among the most abundant (and stable) nuclei in the universe and that lead-208 (^{208}Pb) is the heaviest stable nuclide.

Magic effects can keep unstable nuclides from decaying as rapidly as would otherwise be expected. For example, the nuclides tin-100 (100Sn) and tin-132 (132Sn) are interesting examples of doubly magic isotopes of tin that are unstable; however they represent endpoints beyond which stability drops off rapidly.

In December 2006 hassium-270 (270Hs), with 108 protons and 162 neutrons, was discovered by an international team of scientists led by the Technical University of Munich having the unusually long half-life of 22 seconds. Hassium-270 evidently forms part of an island of stability, and may even be double magic.

Derivation

Magic numbers are typically obtained by empirical studies; however, if the form of the nuclear potential is known then the Schrödinger equation can be solved for the motion of nucleons and energy levels determined. Nuclear shells are said to occur when the separation between energy levels is significantly greater than the local mean separation.

In the shell model for the nucleus, magic numbers are the numbers of nucleons at which a shell is filled. For instance the magic number 8 occurs when $1s_{1/2}$, $1p_{3/2}$, $1p_{1/2}$ energy levels are filled as there is a large energy gap between the $1p_{1/2}$ and the next highest $1d_{5/2}$ energy levels. The empirical values can be reproduced using the classical shell model with a strong spin-orbit interaction.

The atomic analog to nuclear magic numbers are those numbers of electrons leading to discontinuities in the ionization energy. These occur for the noble gases helium, neon, argon, krypton, xenon, radon and ununoctium. Hence, the "atomic magic numbers" are 2, 10, 18, 36, 54, 86 and 118.

In 2007, Jozsef Garai from Florida International University proposed a mathematical formula describing the periodicity of the nucleus in the periodic system based on the tetrahedron.

Recently, an alternative explanation of magic numbers has been given in terms of symmetry considerations. Based on the fractional extension of the standard rotation group, the ground state properties (including the magic numbers) for metallic clusters and nuclei were simultaneously determined analytically. A specific potential term is not necessary in this model.

Modelling Metal Nanoparticle Optical Properties

The interaction and coupling between light and material systems with nanoscale dimensions constitutes a significant portion of the forefront of contemporary materials science. The electronic properties that commonly dominate how a material interacts with electromagnetic radiation can be widely tuned for nanoscale systems through structural properties. Continuing advances in fabrication and synthesis techniques have resulted in the creation of a diverse set of nanoparticle morphologies from a wide range of materials. This work studies the effects of particle morphology on nanoparticles derived from bulk metal and semiconductor components. The unbound conduction electrons in metal nanoparticles dominate the particle's interaction with electromagnetic radiation. Light of an appropriate frequency can excite the collective motion of the conduction electrons – an excitation referred to as a surface plasmon.In bulk metals, the surface plasmons can propagate on the surface of the metal and give rise to effects such as anomalous transmission as well as enable the engineering of mesoscopic metal devices that can focus the propagating surface plasmons to suboptical-wavelength regions with, for example, a sharp tip. In contrast, nanoparticles do not support propagating surface plasmons.The spatial confinement in all dimensions prevents propagation and localizes the surface plasmon excitation to the nanoparticle. Since the particle is metallic, the localized surface plasmon excitation confines the incident electromagnetic energy primarily to the surface of the nanoparticle therefore enabling enhanced interactions between light and matter within close proximity to the nanoparticle surface. Aggregates of multiple metal nanoparticles of diverse shape and size lead to even stronger surface plasmon localization and thus stronger enhancement of light-matter interactions. The enhancement of metal nanoparticle systems through either surface plasmon focusing or localization enables unprecedented sensing abilities [4] where trace amounts of analyte (down to the level of single molecules) placed within close proximity to the metal experience phenomenally enhanced light-matter interactions.

In this manner, very weak interactions such as Raman scattering can be observed at the single molecule level. Nanoparticles derived from semiconductor materials are distinctly different from their metal counterparts as they do not have free conduction electrons. Instead, the electrons are contained in valence band states, and the electronic properties reflect excitation of the valence electrons into conduction band states across an energetic band gap. The spatial confinement of the nanoparticles plays an important role in the energetic size of the band gap, and consequently, the optical properties such as absorption and emission sensitively depend on particle size and shape. Modern synthesis techniques enable very precise tuning of particle morphology, and thus optical and electronic properties.

As a result, semiconductor nanocrystals are exciting materials for applications including light-emitting devices, photovoltaics, ultrasensitive photodectectors, nanoscale light sources, and nanoscale photocatalysts. The morphological sensitivity of the electronic characteristics of nanoparticle systems can be considered a problem as much as it is a tool. For example, not all semiconductor nanocrystals will make ideal photovoltaic materials. Their structural properties must be precisely tuned to optimize optical absorption and subsequent charge separation. Likewise, careful analysis of metal nanoparticle systems is needed, for example, to minimize contaminant background signal in trace analysis applications or to maximize emission stability in biological labelling applications. In this work, careful consideration is given to the structural properties of both single metal nanoparticle aggregates and single semiconductor nanoparticles with the goal of identifying crucial properties and characteristics that should be optimized for a variety of practical applications. This introductory section serves to briefly introduce the electronic and optical properties of both metal nanoparticle aggregates and semiconductor nanoparticles. The difference between bulk material electronic properties and those of the corresponding nanoparticles is briefly discussed for both metal and semiconductor systems. Then, an introduction to metal nanoparticle aggregates is approached in a bottom- up manner where the electronic properties of the nanoparticle building blocks are established before discussing their interaction in an aggregate system which gives rise to surface-enhanced optical processes.

Next, a brief discussion of the structural and electronic properties of colloidal semiconductor nanoparticles is given. The introduction is then concluded with overviews of both the unique properties of single quantum emitters and the nonlinear optical response of materials. Following the introduction, Electronic properties of nanoparticles: from clusters to nanoparticles In general, "cluster science" refers to the study of material systems with physical dimensions which are comparable to those of the constituent atoms. At these length scales, which range from 10 Ú A to over 100 nm, the material systems are discrete particles consisting of as few as 2 atoms in molecular clusters to more than 10 7 atoms in larger nanoparticles. In this size regime, material systems are in an electronic configuration which bridges discrete molecular-like electronic states and the bulk electronic bands which are associated with the material's familiar macroscopic properties.

For both metals and semiconductors, the fundamental building blocks of the material are the constituent atoms in the primitive cells of the crystalline lattice. In general, the primitive cell may contain multiple atoms, but for the sake of clarity, the discussion here focuses on single-atom materials. As two atoms are brought together to form a dimer, the molecular states become an energetically split symmetric and antisymmetric combination of the individual atomic states, and the highest occupied and lowest unoccupied atomic orbitals become the highest occupied and the lowest unoccupied molecular orbitals (HOMO and LUMO, respectively). In an identical manner, further splitting occurs in larger clusters with more atoms. Eventually in clusters with a relatively large number of atoms, the splitting becomes negligible (i.e.,less than the thermal energy), and the set of closely separated states centred around the original atomic energy level can be considered a quasicontinuous band. The evolution of the electronic energy levels of metals and semiconductors as particle size increases from individual atoms to the bulk to the electronic configuration of the corresponding cluster or bulk material. A band of states forms for each electronic orbital of the single atom that makes up the constituent building blocks. While the energetic centres of the bands are determined by the atomic orbitals, the number of states within the bands are determined by the total number of atoms in the system:each atom "contributes" a single state to the band which can contain two electrons of opposite spin. Therefore, an atomic orbital which contains and even number of valence electrons leads to a filled electronic band in the bulk whereas an odd number of valence electrons leads to a half-filled electronic band. Therein lies the fundamental difference between semiconductors and metals. In metals, where the highest energy individual atomic orbital is filled with an odd number of valence electrons, the highest energy electronic band is only half-filled. This half-filled band – called the conduction band – supports electrical conduction due to the nearby, higher energy, vacant states within the top half of the band. In contrast, the highest occupied orbital in semiconductor atoms is filled with an even number of valence electrons leading to complete filling of the top-most band – called the valence band.

Thereby, the closest vacant state for semiconductor electrons lies in the next higher band – also referred to as the conduction band – and thus interband excitation of the electrons is required for significant electrical conductivity. The difference in occupancy of the highest-filled band between semiconductors and metals has important consequences for their corresponding nanoparticle systems. The amount of splitting between the states in a cluster is not equal ; the states in the middle of the quasiband are less separated than those at the edges. Thus, the position of the highest occupied electronic state within this quasiband plays an important role in determining the critical particle size at which a material makes the molecular-to- bulk transition. For metals, where the highest occupied electronic level lies in the middle of the quasiband, nanoparticles assume bulk-like properties at smaller sizes than their semiconductor counterparts in which the highest occupied state lies at the edge of the quasiband. So, while metal clusters quickly become conducting with increasing size, semiconductor nanoparticles preserve molecular-like behaviour of their highest occupied electronic states to notably larger particle dimensions. As a 6 consequence, the task of deliberately tuning the intrinsic electronic states through fabrication or synthesis is much easier for nanoparticles derived from semiconductor materials as compared to metals. This work deals with metal and semiconductor nanoscale systems on length scales which span both sides of the molecular-to-bulk transition. For the sake of clarity, the term" cluster" is explicitly used for material systems where it is assumed that discrete energy states exist and energy bands are not present at the relevant temperature. Otherwise, the less-specific term "nanoparticle" is used for material systems where discrete states are not assumed. 1.2 Metal nanoparticles and rough films.

A large portion of this work studies the intrinsic optical response of semi-continuous silver films. The physical structure of these metal films consists of an aggregation of metal clusters and nanoparticles ranging from two-atom dimers to particles several-hundred nanometres in size. Illumination with visible or infrared radiation excites each of these constituent building blocks and the complex set of interactions within the film geometry give rise to an elaborate optical response. Here, a bottom-up approach is used to first describe the optical and electronic properties of the individual building blocks, and then to establish a qualitative understanding of how their interactions in aggregate structures can give rise to the rich optical properties of a simple metal film. The optical response of metallic nanoparticles and clusters is extraordinarily sen- sitive to particle scale over a wide range of sizes. In addition to the aforementioned transition from molecular-like electronic states to bulk-like bands at small length scales, the kinetics and electrodynamic response of the nearly-free electrons in the conduction band exhibit their own unique size effects. The first category contains small metal clusters with discrete, molecular-like electronic states. The electronic states become a quasi continuous band for metal nanoparticles exceeding 2 nm in diameter.

8

Synthesis and Characterization of Nano-Manganese Dioxide

Manganese(IV) oxide is the inorganic compound with the formula MnO2. This blackish or brown solid occurs naturally as the mineral pyrolusite, which is the main ore of manganese and a component of manganese nodules. The principal use for MnO2 is for dry-cell batteries, such as the alkaline battery and the zinc-carbon battery. MnO2 is also used as a pigment and as a precursor to other manganese compounds, such as KMnO4. It is used as a reagent in organic synthesis, for example, for the oxidation of allylic alcohols.

Structure

Several polymorphs of MnO2 are claimed, as well as a hydrated form. Like many other dioxides, MnO2 crystallizes in the rutile crystal structure (this polymorph is called β-MnO2), with three-coordinate oxide and octahedral metal centres. MnO2 is characteristically nonstoichiometric, being deficient in oxygen. The complicated solid state chemistry of this material is relevant to the lore of "freshly prepared" MnO2 in organic synthesis.

Production

Naturally occurring manganese dioxide contains impurities and a considerable amount of manganese in its 3+ oxidation state. Only a limited number of deposits contain the γ modification in purity sufficient for the battery industry. The production of ferrite also requires high purity manganese dioxide. Therefore the production of synthetic manganese dioxide is important. Two groups of methods are used, yielding "chemical manganese dioxide" (CMD) and "electrolytical manganese dioxide" (EMD). The CMD is mostly used for the production of ferrites, whereas EMD is used for the production of batteries.

Chemical Manganese Dioxide

One of the two chemical methods starts from natural manganese dioxide and converts it using dinitrogen tetroxide and water to manganese(II) nitrate solution. Evaporation of the water, leaves the crystalline nitrate salt. At temperatures of 400 °C, the salt decomposes, releasing N_2O_4 and leaving a residue of purified manganese dioxide. These two steps can be summarized as:

$$MnO_2 + N_2O_4 \rightleftarrows Mn(NO_3)_2$$

In the other chemical process, manganese dioxide ore is reduced by heating with oil or coal. The resulting manganese(II) oxide is dissolved in sulfuric acid, and the filtered solution is treated with ammonium carbonate to precipitate $MnCO_3$. The carbonate is calcined in air to give a mixture of manganese(II) and manganese(IV) oxides. To complete the process, a suspension of this material in sulfuric acid is treated with sodium chlorate. Chloric acid, which forms in situ, converts any Mn(III) and Mn(II) oxides to the dioxide, releasing chlorine as a by-product.

Electrolytical Manganese Dioxide

Electrolytical manganese dioxide (EMD) is used in zinc-carbon batteries together with zinc chloride and ammonium chloride. EMD is commonly used in zinc manganese dioxide rechargeable alkaline (Zn RAM) cells also. For these applications, purity is extremely important.

Reactions

The important reactions of MnO_2 are associated with its redox, both oxidation and reduction.

Reduction

MnO_2 is the principal precursor to ferromanganese and related alloys, which are widely used in the steel industry. The conversions involve carbothermal reduction using coke:

$$MnO2 + 2\ C \rightarrow Mn + 2\ CO$$

The key reactions of MnO_2 in batteries is the one-electron reduction:

$$MnO_2 + e^- + H^+ \rightarrow MnO(OH)$$

MnO2 catalyses several reactions that form O_2. In a classical laboratory demonstration, heating a mixture of potassium chlorate and manganese dioxide produces oxygen gas. Manganese dioxide also catalyses the decomposition of hydrogen peroxide to oxygen and water:

$$2\ H_2O_2 \rightarrow 2\ H_2O + O_2$$

Manganese dioxide decomposes above about 530 °C to manganese(III) oxide and oxygen. At temperatures close to 1000 °C, the mixed-valence compound Mn_3O_4 forms. Higher temperatures give MnO.

Hot concentrated sulfuric acid reduces the MnO_2 to manganese(II) sulfate:

$$2\ MnO_2 + 2\ H_2SO_4 \rightarrow 2\ MnSO_4 + O_2 + 2\ H_2O$$

The reaction of hydrogen chloride with MnO_2 was used by Carl Wilhelm Scheele in the original isolation of chlorine gas in 1774:

$$MnO_2 + 4\ HCl \rightarrow MnCl_2 + Cl_2 + 2\ H_2O$$

As a source of hydrogen chloride, Scheele treated sodium chloride with concentrated sulfuric acid.

$$E^\circ\ (MnO_2(s) + 4\ H^+ + 2\ e^- \rightleftharpoons Mn^{2+} + 2\ H_2O) = +1.23\ V$$

$$E^\circ\ (Cl_2(g) + 2\ e^- \rightleftharpoons 2\ Cl^-) = +1.36\ V$$

The standard electrode potentials for the half reactions indicate that the reaction is endothermic at pH = 0 (1 M [H^+]), but it is favoured by the lower pH as well as the evolution (and removal) of gaseous chlorine.

This reaction is also a convenient way to remove the manganese dioxide precipitate from the ground glass joints after running a reaction (i. e., an oxidation with potassium permanganate).

Oxidation

Heating a mixture of KOH and MnO_2 in air gives green potassium manganate:

$$2\ MnO_2 + 4\ KOH + O_2 \rightarrow 2\ K_2MnO_4 + 2\ H_2O$$

Potassium manganate is the precursor to potassium permanganate, a common oxidant.

Applications

The predominant application of MnO_2 is as a component of dry cell batteries, so called Leclanché cell, or zinc–carbon batteries. Approximately 500,000 tonnes are consumed for this application annually.

Other industrial applications include the use of MnO2 as an inorganic pigment in ceramics and in glassmaking.

Organic Synthesis

A specialised use of manganese dioxide is as oxidant in organic synthesis. The effectiveness of the reagent depends on the method of preparation, a problem that is typical for other heterogeneous reagents where surface area, among other variables, is a significant factor. The mineral pyrolusite makes a poor reagent. Usually, however, the reagent is generated in situ by treatment of an aqueous solution $KMnO_4$ with a Mn(II) salt, typically the sulfate. MnO_2 oxidizes allylic alcohols to the corresponding aldehydes or ketones:

$$\text{cis-RCH=CHCH}_2\text{OH} + \text{MnO}_2 \rightarrow \text{cis-RCH=CHCHO} + \text{“MnO”} + \text{H}_2\text{O}$$

The configuration of the double bond is conserved in the reaction. The corresponding acetylenic alcohols are also suitable substrates, although the resulting propargylic aldehydes can be quite reactive. Benzylic and even unactivated alcohols are also good substrates. 1,2-Diols are cleaved by MnO_2 to dialdehydes or diketones. Otherwise, the applications of MnO_2 are numerous, being applicable to many kinds of reactions including amine oxidation, aromatization, oxidative coupling, and thiol oxidation.

Pigment

Manganese dioxide was one of the earliest natural substances used by human ancestors. It was used as a pigment at least from the middle paleolithic. It was possibly used first for body painting, and later for cave painting. Some of the most famous early cave paintings in Europe were executed by means of manganese dioxide.

Hazards

Manganese dioxide can slightly stain human skin if it is damp or in a heterogeneous mixture, but the stains can be washed off quite easily with some rubbing.

Nanostructured Mn Oxide

Nanostructured Mn oxide clusters supported on mesoporous silica KIT-6 efficiently evolve O2 in aqueous solution under mild conditions.

Generation of a solar fuel on a terawatt scale requires that water is used as electron source.

Catalysts for the four-electron oxidation of water to oxygen need not only to fulfill the rate and size requirements for keeping up with the solar flux, but must be robust and made of earth abundant elements.

Benefiting from explorations of first row transition metal oxides as O2 evolving electrocatalysts and, in a few cases, chemically driven catalysts over the past several decades, progress on Co oxide-based water oxidation catalysts has recently been reported on two fronts. Nocera's group has employed electrodeposition of Co-containing films on an indium tin oxide anode from an aqueous Co phosphate solution and achieved O2 evolution at neutral pH and with modest overpotential (10 mA current, 100 percent Faradaic efficiency, 410 mV overpotential).

In another direction, a nanostructured silica support has enabled the synthesis of nanometre-sized Co3O4 (spinel) clusters which possess a very high surface area (bundles of nanorods of 8 nm diameter).

A recent detailed study by Tilley and Bell et al. of the size-dependence of the

catalytic water oxidation activity for cubic Co3O4 nano-particles (range 5 to 50 nm) loaded on Ni foam anodes in alkaline aqueous solution showed a clear, linear dependence of surface area thus confirming the surface size effect.

In light of the fact that Mn is not only earth abundant but also environmentally friendly, this metal is particularly attractive for use as a catalyst material. In fact, electrochemists have conducted extensive studies on the use of Mn oxides as anode coatings for catalytic O2evolution from water.

While work was typically conducted under basic conditions around pH 14 to minimize the overvoltage, Tamura et al. studied water oxidation over a wide range of pH values including neutral aqueous solution for which they reported current densities around 100 mA cm corresponding to a TOF of Z 0.013 s1 at an overvoltage of 440 mV (the TOF is a lower limit because no BET surface area was reported; therefore, the estimate assumes that all deposited Mn centres are catalytically active. Also under mild conditions (pH 5), Harriman et al. demonstrated oxygen evolution in aqueous suspensions of mm-sized Mn2O3 particles using photochemically generated

Ru3+(bpy)3 as oxidant for driving the catalyst.

Taken together, these literature reports suggest that Mn oxides hold promise for developing robust, efficient water oxidation catalysts that can be operated under mild pH and temperature conditions if nano-sized clusters with high surface area can be made.+

Synthesis and Characterization of Zno Nanoparticles

Zinc oxide is an inorganic compound with the formula ZnO. ZnO is a white powder that is insoluble in water, and it is widely used as an additive in numerous materials and products including rubbers, plastics, ceramics, glass, cement, lubricants, paints, ointments, adhesives, sealants, pigments, foods (source of Zn nutrient), batteries, ferrites, fire retardants, and first-aid tapes. It occurs naturally as the mineral zincite, but most zinc oxide is produced synthetically.

In materials science, ZnO is a wide-bandgap semiconductor of the II-VI semiconductor group (since oxygen was classed as an element of VIA group (the 6th main group, now referred to as 16th) of the periodic table and zinc, a transition metal, as a member of the IIB (2nd B), now 12th, group).

The native doping of the semiconductor (due to oxygen vacancies or zinc interstitials) is n-type. This semiconductor has several favourable properties, including good transparency, high electron mobility, wide bandgap, and strong room-temperature luminescence. Those properties are used in emerging applications for transparent electrodes in liquid crystal displays, in energy-saving or heat-protecting windows, and in electronics as thin-film transistors and light-emitting diodes.

Chemical Properties

ZnO occurs as a white powder. The mineral zincite usually contains manganese and other impurities that confer a yellow to red colour.

Crystalline zinc oxide is thermochromic, changing from white to yellow when heated and in air reverting to white on cooling. This colour change is caused by a small loss of oxygen to the environment at high temperatures to form the non-stoichiometric $Zn_{1+x}O$, where at 800 °C, x = 0.00007.

Zinc oxide is an amphoteric oxide. It is nearly insoluble in water, but it is soluble in (degraded by) most acids, such as hydrochloric acid:

$$ZnO + 2\ HCl \rightarrow ZnCl_2 + H_2O$$

Bases also degrade the solid to give soluble zincates:

$$ZnO + 2\ NaOH + H_2O \rightarrow Na_2[Zn(OH)_4]$$

ZnO reacts slowly with fatty acids in oils to produce the corresponding carboxylates, such as oleate or stearate. ZnO forms cement-like products when mixed with a strong aqueous solution of zinc chloride and these are best described as zinc hydroxy chlorides. This cement was used in dentistry.

Hopeite

ZnO also forms cement-like material when treated with phosphoric acid; related materials are used in dentistry. A major component of zinc phosphate cement produced by this reaction is hopeite, $Zn_3(PO_4)_2 \cdot 4H_2O$.

ZnO decomposes into zinc vapor and oxygen at around 1975 °C with a standard oxygen pressure. In a carbothermic reaction, heating with carbon converts the oxide into zinc vapor at a much lower temperature (around 950 °C).

$$ZnO + C \rightarrow Zn_{(Vapor)} + CO$$

Zinc oxide can react violently with aluminium and magnesium powders, with chlorinated rubber and linseed oil on heating causing fire and explosion hazard.

It reacts with hydrogen sulfide to give the sulfide. This reaction is used commercially.

$$ZnO + H_2S \rightarrow ZnS + H_2O$$

Structure

Zinc oxide crystallizes in two main forms, hexagonal wurtzite and cubic zincblende. The wurtzite structure is most stable at ambient conditions and thus most common.

The zincblende form can be stabilized by growing ZnO on substrates with cubic lattice structure. In both cases, the zinc and oxide centres are tetrahedral, the most characteristic geometry for Zn(II). ZnO converts to the rocksalt motif at relatively high pressures about 10 GPa.

Hexagonal and zincblende polymorphs have no inversion symmetry (reflection of a crystal relative to any given point does not transform it into itself). This and other lattice symmetry properties result in piezoelectricity of the hexagonal and zincblende ZnO, and pyroelectricity of hexagonal ZnO.

The hexagonal structure has a point group 6 mm (Hermann-Mauguin notation) or C_{6v} (Schoenflies notation), and the space group is $P6_3mc$ or $C_{6v}{}^4$. The lattice constants are a = 3.25 Å and c = 5.2 Å; their ratio c/a ~ 1.60 is close to the ideal value for hexagonal cell c/a = 1.633. As in most group II-VI materials, the bonding in ZnO is largely ionic (Zn^{2+}–O^{2-}) with the corresponding radii of 0.074 nm for Zn^{2+} and 0.140 nm for O^{2-}.

This property accounts for the preferential formation of wurtzite rather than zinc blende structure, as well as the strong piezoelectricity of ZnO. Because of the polar Zn-O bonds, zinc and oxygen planes are electrically charged. To maintain electrical neutrality, those planes reconstruct at atomic level in most relative materials, but not in ZnO – its surfaces are atomically flat, stable and exhibit no reconstruction. This anomaly of ZnO is not fully explained yet.

Mechanical Properties

ZnO is a relatively soft material with approximate hardness of 4.5 on the Mohs scale. Its elastic constants are smaller than those of relevant III-V semiconductors, such as GaN. The high heat capacity and heat conductivity, low thermal expansion and high melting temperature of ZnO are beneficial for ceramics. ZnO exhibits a very long lived optical phonon E2(low) with a lifetime as high as 133 ps at 10 K.

Among the tetrahedrally bonded semiconductors, it has been stated that ZnO has the highest piezoelectric tensor, or at least one comparable to that of GaN and AlN. This property makes it a technologically important material for many piezoelectrical applications, which require a large electromechanical coupling.

Electrical Properties

ZnO has a relatively large direct band gap of ~3.3 eV at room temperature. Advantages associated with a large band gap include higher breakdown voltages, ability to sustain large electric fields, lower electronic noise, and high-temperature and high-power operation. The bandgap of ZnO can further be tuned to ~3–4 eV by its alloying with magnesium oxide or cadmium oxide. Most ZnO has *n*-type character, even in the absence of intentional doping. Nonstoichiometry is typically the origin of n-type character, but the subject remains controversial. An alternative explanation has been proposed, based on theoretical calculations, that unintentional substitutional hydrogen impurities are responsible. Controllable n-type doping is easily achieved by substituting Zn with group-III elements such as Al, Ga, In or by substituting oxygen with group-VII elements chlorine or iodine.

Reliable p-type doping of ZnO remains difficult. This problem originates from low solubility of p-type dopants and their compensation by abundant n-type impurities. This problem is observed with GaN and ZnSe. Measurement of p-type in "intrinsically" n-type material is complicated by the inhomogeneity of samples.

Current limitations to p-doping does not limit electronic and optoelectronic applications of ZnO, which usually require junctions of n-type and p-type material. Known p-type dopants include group-I elements Li, Na, K; group-V elements N, P and As; as well as copper and silver. However, many of these form deep acceptors and do not produce significant p-type conduction at room temperature.

Electron mobility of ZnO strongly varies with temperature and has a maximum of ~2000 $cm^2/(V \cdot s)$ at 80 K. Data on hole mobility are scarce with values in the range 5–30 $cm^2/(V \cdot s)$.

Production

For industrial use, ZnO is produced at levels of 10^5 tons per year by three main processes:

Indirect Process

In the indirect or French process, metallic zinc is melted in a graphite crucible and vaporized at temperatures above 907 °C (typically around 1000 °C). Zinc vapor reacts with the oxygen in the air to give ZnO, accompanied by a drop in its temperature and bright luminescence. Zinc oxide particles are transported into a cooling duct and collected in a bag house. This indirect method was popularized by LeClaire (France) in 1844 and therefore is commonly known as the French process. Its product normally consists of agglomerated zinc oxide particles with an average size of 0.1 to a few micrometres. By weight, most of the world's zinc oxide is manufactured via French process.

Direct Process

The direct or American process starts with diverse contaminated zinc composites, such as zinc ores or smelter by-products. The zinc precursors are reduced (carbothermal reduction) by heating with a source of carbon such as anthracite to produce zinc vapor, which is then oxidized as in the indirect process. Because of the lower purity of the source material, the final product is also of lower quality in the direct process as compared to the indirect one.

Wet Chemical Process

A small amount of industrial production involves wet chemical processes, which start with aqueous solutions of purified zinc salts, from which zinc carbonate or zinc hydroxide is precipitated. The precipitate is then filtered, washed, dried and calcined at temperatures around 800 °C.

Laboratory Synthesis

Figure: *Synthetic ZnO crystals. Red and green colour are associated with different concentrations of oxygen vacancies.*

A large number of specialised methods exist for producing ZnO for scientific studies and niche applications. These methods can be classified by the resulting ZnO form (bulk, thin film, nanowire), temperature ("low", that is close to room temperature or "high", that is T ~ 1000 °C), process type (vapor deposition or growth from solution) and other parameters.

Large single crystals (many cubic centimetres) can be grown by the gas transport (vapor-phase deposition), hydrothermal synthesis, or melt growth. However, because of high vapor pressure of ZnO, growth from the melt is problematic. Growth by gas transport is difficult to control, leaving the hydrothermal method as a preference. Thin films can be produced by chemical vapor deposition, metalorganic vapour phase epitaxy, electrodeposition, pulsed laser deposition, sputtering, sol-gel synthesis, atomic layer deposition, spray pyrolysis, etc.

Ordinary white powdered zinc oxide can be produced in the laboratory by electrolyzing a solution of sodium bicarbonate with a zinc anode. Zinc hydroxide and hydrogen gas are produced. The zinc hydroxide upon heating decomposes to zinc oxide.

$$Zn + 2\ H_2O \rightarrow Zn(OH)_2 + H_2$$

$$Zn(OH)_2 \rightarrow ZnO + H_2O$$

ZnO Nanostructures

Nanostructures of ZnO can be synthesized into a variety of morphologies including nanowires, nanorods, tetrapods, nanobelts, nanoflowers, nanoparticles etc. Nanostructures can be obtained with most above-mentioned techniques, at certain conditions, and also with the vapor-liquid-solid method.

Rodlike nanostructures of ZnO can be produced via aqueous methods, which are attractive for the following reasons: They are low cost, less hazardous, and thus capable of easy scaling up; the growth occurs at a relatively low temperature, compatible with flexible organic substrates; there is no need for the use of metal catalysts, and thus it can be integrated with well-developed silicon technologies. In addition, there are a variety of parameters that can be tuned to effectively control the morphology and properties of the final product. Wet chemical methods have been demonstrated as a very powerful and versatile technique for growing one-dimensional ZnO nanostructures. The synthesis is typically carried out at temperatures of about 90 °C, in an equimolar aqueous solution of zinc nitrate and hexamine, the latter providing the basic environment. Certain additives, such as polyethylene glycol or polyethylenimine, can improve the aspect ratio of the ZnO nanowires.

Doping of the ZnO nanowires has been achieved by adding other metal nitrates to the growth solution. The morphology of the resulting nanostructures can be tuned by changing the parameters relating to the precursor composition (such as the zinc concentration and pH) or to the thermal treatment (such as the temperature and heating rate).

Aligned ZnO nanowires on pre-seeded silicon, glass and gallium nitride substrates have been grown in aqueous solutions using aqueous zinc salts such as Zinc nitrate and Zinc acetate in basic environments. Pre-seeding substrates with ZnO creates sites for homogeneous nucleation of ZnO crystal during the synthesis. Common pre-seeding methods include in-situ thermal decomposition of zinc acetate crystallites, spincoating of ZnO nanoparticles and the use of physical vapor deposition methods to deposit ZnO thin films. Pre-seeding can be performed in conjunction with top down patterning methods such as electron beam lithography and nanosphere lithography to designate nucleation sites prior to growth. Aligned ZnO nanowires can be used in dye-sensitized solar cells and field emission devices.

History

Zinc compounds were probably used by early humans, in processed and unprocessed forms, as a paint or medicinal ointment, but their composition is uncertain. The use of *pushpanjan*, probably zinc oxide, as a salve for eyes and open wounds, is mentioned in the Indian medical text the Charaka Samhita, thought to date from 500 BC or before. Zinc oxide ointment is also mentioned by the Greek physician Dioscorides (1st century AD.) Avicenna mentions zinc oxide in *The Canon of Medicine* (1025 AD), which mentioned it as a preferred treatment for a variety of skin conditions, including skin cancer. Though it is no longer used for treating skin cancer, it is still widely used to treat a variety of other skin conditions, in products such as baby powder and creams against diaper rashes, calamine cream, anti-dandruff shampoos, and antiseptic ointments.

The Romans produced considerable quantities of brass (an alloy of zinc and copper) as early as 200 BC by a cementation process where copper was reacted with zinc oxide. The zinc oxide is thought to have been produced by heating zinc ore in a shaft furnace. This liberated metallic zinc as a vapor, which then ascended the flue and condensed as the oxide. This process was described by Dioscorides in the 1st century AD. Zinc oxide has also been recovered from zinc mines at Zawar in India, dating from the second half of the first millennium BC. This was presumably also made in the same way and used to produce brass.

From the 12th to the 16th century zinc and zinc oxide were recognised and produced in India using a primitive form of the direct synthesis process. From India, zinc manufacture moved to China in the 17th century. In 1743, the first European zinc smelter was established in Bristol, United Kingdom.

The main usage of zinc oxide (zinc white) was again paints and additive to ointments. Zinc white was accepted as a watercolour by 1834 but it did not mix well with oil. This problem was quickly solved by optimizing the synthesis of ZnO. In 1845, LeClaire in Paris was producing the oil paint on a large scale, and by 1850, zinc white was being manufactured throughout Europe.

The success of zinc white paint was due to its advantages over the traditional white lead: zinc white is essentially permanent in sunlight, it is not blackened by sulfur-bearing air, it is non-toxic and more economical. Because zinc white is so "clean" it is very valuable for making tints with other colours; however, it makes a rather brittle dry film when unmixed with other colours.

For example, during the late 1890s and early 1900s, some artists used zinc white as a ground for their oil paintings. All those paintings developed cracks over the years.

In the recent times, most zinc oxide was used in the rubber industry. In the 1970s, the second largest application of ZnO was photocopying. High-quality ZnO produced by the "French process" was added into the photocopying paper as a filler. This application was however soon displaced.

Applications

The applications of zinc oxide powder are numerous, and the principal ones are summarized below. Most applications exploit the reactivity of the oxide as a precursor to other zinc compounds. For material science applications, zinc oxide has high refractive index, high thermal conductivity, binding, antibacterial and UV-protection properties. Consequently, it is added into materials and products including plastics, ceramics, glass, cement, rubber, lubricants, paints, ointments, adhesive, sealants, concrete manufacturing, pigments, foods, batteries, ferrites, fire retardants, etc.

Rubber Manufacture

Between 50% and 60% of ZnO use is in the rubber industry. Zinc oxide along with stearic acid is used in the vulcanization of rubber ZnO additive also protect rubber from fungi and UV light.

Ceramic Industry

Ceramic industry consumes a significant amount of zinc oxide, in particular in ceramic glaze and frit compositions. The relatively high heat capacity, thermal conductivity and high temperature stability of ZnO coupled with a comparatively low coefficient of expansion are desirable properties in the production of ceramics. ZnO affects the melting point and optical properties of the glazes, enamels, and ceramic formulations. Zinc oxide as a low expansion, secondary flux improves the elasticity of glazes by reducing the change in viscosity as a function of temperature and helps prevent crazing and shivering. By substituting ZnO for BaO and PbO, the heat capacity is decreased and the thermal conductivity is increased. Zinc in small amounts improves the development of glossy and brilliant surfaces. However in moderate to high amounts, it produces matte and crystalline surfaces. With regard to colour, zinc has a complicated influence.

Medicine

Zinc oxide as a mixture with about 0.5% iron(III) oxide (Fe_2O_3) is called calamine and is used in calamine lotion. Two minerals, zincite and hemimorphite, have been historically called calamine. When mixed with eugenol, a ligand, zinc oxide eugenol is formed, which has applications as a restorative and prosthodontic in dentistry.

Reflecting the basic properties of ZnO, fine particles of the oxide have deodorizing and antibacterial properties and for that reason are added into materials including cotton fabric, rubber, oral care products, and food packaging. Enhanced antibacterial action of fine particles compared to bulk material is not exclusive to ZnO and is observed for other materials, such as silver. This property is due to the increased surface area of the fine particles.

Zinc oxide is widely used to treat a variety of other skin conditions, in products such as baby powder and barrier creams to treat diaper rashes, calamine cream, anti-dandruff shampoos, and antiseptic ointments. It is also a component in tape (called "zinc oxide tape") used by athletes as a bandage to prevent soft tissue damage during workouts.

Zinc oxide can be used in ointments, creams, and lotions to protect against sunburn and other damage to the skin caused by ultraviolet light. It is the broadest spectrum UVA and UVB reflector that is approved for use as a sunscreen by the FDA, and is completely photostable. When used as an ingredient in sunscreen, zinc oxide blocks both UVA (320–400 nm) and UVB (280–320 nm) rays of ultraviolet light. Zinc oxide and the other most common physical sunscreen, titanium dioxide, are considered to be nonirritating, nonallergenic, and non-comedogenic. Zinc oxide is, however, slightly absorbed into the skin

Many sunscreens use nanoparticles of zinc oxide (along with nanoparticles of titanium dioxide) because such small particles do not scatter light and therefore do not appear white. There has been concern that they might be absorbed into the skin, and a study published in 2010 found that nanoparticles of ZnO that were applied to human skin via sunscreens could be traced in venous blood and urine samples. In contrast, a comprehensive review of the medical literature from 2011 says that no evidence of systemic absorption can be found in the literature.

Zinc oxide nanoparticles can enhance the antibacterial activity of Ciprofloxacin. It has been shown that nano ZnO which has the average size between 20 nm and 45 nm can enhance the antibacterial activity of Ciprofloxacin against *Staphylococcus aureus* and *Escherichia coli* in Vitro. The enhancing effect of this nanomaterial is concentration-dependent against all test strains.

This effect may be due to two reasons. First, Zinc Oxide nanoparticles can interfere with NorA protein. NorA is a protein which is developed for conferring resistance in bacteria and has pumping activity that mediate the effluxing of hydrophilic fluroquinolones from a cell. Second, Zinc Oxide nanoparticles can interfere with Omf protein. Omf is a membrane protein that is responsible for the permeation of quinolones into the cell.

Cigarette Filters

Zinc oxide is a constituent of cigarette filters. A filter consisting of charcoal impregnated with zinc oxide and iron oxide removes significant amounts of HCN and H_2S from tobacco smoke without affecting its flavor.

Food Additive

Zinc oxide is added to many food products, including breakfast cereals, as a source of zinc, a necessary nutrient. (Zinc sulfate is also used for the same purpose.) Some prepackaged foods also include trace amounts of ZnO even if it is not intended as a nutrient.

Pigment

Zinc white is used as a pigment in paints and is more opaque than lithopone, but less opaque than titanium dioxide. It is also used in coatings for paper. Chinese white is a special grade of zinc white used in artists' pigments. It is also a main ingredient of mineral makeup.

UV Absorber

Micronized and nano-scale zinc oxide and titanium dioxide provide strong protection against UVA ultraviolet radiation, and are used in suntan lotion, and also in UV-blocking sunglasses for use in space and for protection when welding, following research by scientists at JPL.

Coatings

Paints containing zinc oxide powder have long been utilised as anticorrosive coatings for metals. They are especially effective for galvanized iron. Iron is difficult to protect because its reactivity with organic coatings leads to brittleness and lack of adhesion. Zinc oxide paints retain their flexibility and adherence on such surfaces for many years.

ZnO highly n-type doped with Al, Ga, or In is transparent and conductive (transparency ~90%, lowest resistivity ~10^{-4} $\Omega \cdot$cm). ZnO:Al coatings are used for energy-saving or heat-protecting windows. The coating lets the visible part of the spectrum in but either reflects the infrared (IR) radiation back into the room (energy saving) or does not let the IR radiation into the room (heat protection), depending on which side of the window has the coating.

Plastics, such as polyethylene naphthalate (PEN), can be protected by applying zinc oxide coating. The coating reduces the diffusion of oxygen with PEN. Zinc oxide layers can also be used on polycarbonate (PC) in outdoor applications. The coating protects PC from solar radiation and decreases the oxidation rate and photo-yellowing of PC.

Corrosion Prevention in Nuclear Reactors

Zinc oxide depleted in the zinc isotope with the atomic mass 64 is used in corrosion prevention in nuclear pressurized water reactors. The depletion is necessary, because ^{64}Zn is transformed into radioactive ^{65}Zn under irradiation by the reactor neutrons.

Methane Reforming

Zinc oxide (ZnO) is used as a pretreatment step to remove hydrogen sulfide (H_2S) from natural gas following hydrogenation of any sulfur compounds prior to a methane reformer, which can poison the catalyst. At temperatures between about 230–430 °C (446–806 °F), H_2S is converted to water by the following reaction:

$$H_2S + ZnO \rightarrow H_2O + ZnS$$

The zinc sulfide (ZnS) is replaced with fresh zinc oxide when the zinc oxide has been consumed.

Potential Applications

Electronics

ZnO has wide direct band gap (3.37 eV or 375 nm at room temperature). Therefore, its most common potential applications are in laser diodes and light emitting diodes (LEDs). Some optoelectronic applications of ZnO overlap with that of GaN, which has a similar bandgap (~3.4 eV at room temperature). Compared to GaN, ZnO has a larger exciton binding energy (~60 meV, 2.4 times of the room-temperature thermal energy), which results in bright room-temperature emission from ZnO. ZnO can be combined with GaN for LED-applications. For instance as TCO layer and ZnO nanostructures provide better light outcoupling. Other properties of ZnO favourable for electronic applications include its stability to high-energy radiation and to wet chemical etching. Radiation resistance makes ZnO a suitable candidate for space applications. ZnO is the most promising candidate in the field of random lasers to produce an electronically pumped UV laser source.

The pointed tips of ZnO nanorods result in a strong enhancement of an electric field. Therefore, they can be used as field emitters.

Aluminium-doped ZnO layers are used as a transparent electrodes. The constituents Zn and Al are much cheaper and less toxic compared to the generally used indium tin oxide (ITO). One application which has begun to be commercially available is the use of ZnO as the front contact for solar cells or of liquid crystal displays.

Transparent thin-film transistors (TTFT) can be produced with ZnO. As field-effect transistors, they even may not need a p–n junction, thus avoiding the p-type doping problem of ZnO. Some of the field-effect transistors even use ZnO nanorods as conducting channels.

Zinc Oxide Nanorod Sensor

Zinc oxide nanorod sensors are devices detecting changes in electrical current passing through zinc oxide nanowires due to adsorption of gas molecules. Selectivity to hydrogen gas was achieved by sputtering Pd clusters on the nanorod surface. The addition of Pd appears to be effective in the catalytic dissociation of hydrogen molecules into atomic hydrogen, increasing the sensitivity of the sensor device. The sensor detects hydrogen concentrations down to 10 parts per million at room temperature, whereas there is no response to oxygen.

Spintronics

ZnO has also been considered for spintronics applications: if doped with 1–10% of magnetic ions (Mn, Fe, Co, V, etc.), ZnO could become ferromagnetic, even at room temperature. Such room temperature ferromagnetism in ZnO:Mn has been observed, but it is not clear yet whether it originates from the matrix itself or from secondary oxide phases.

Piezoelectricity

The piezoelectricity in textile fibres coated in ZnO have been shown capable of fabricating "self-powered nanosystems" with everyday mechanical stress from wind or body movements.

In 2008 the *Centre for Nanostructure Characterization* at the Georgia Institute of Technology reported producing an electricity generating device (called flexible charge pump generator) delivering alternating current by stretching and releasing zinc oxide nanowires. This mini-generator creates an oscillating voltage up to 45 millivolts, converting close to seven percent of the applied mechanical energy into electricity. Researchers used wires with lengths of 0.2–0.3 mm and diameters of three to five micrometres, but the device could be scaled down to smaller size.

Zinc Oxide (ZnO) Nanoparticles – Properties, Applications

Nanotechnology research has gained momentum in the recent years by providing innovative solutions in the field of biomedical, materials science, optics and electronics. Nanoparticles are essentially a varied form of basic elements derived by altering their atomic and molecular properties of elements. This article elaborates on the properties and applications of zinc oxide nanoparticles.

Zinc oxide (ZnO) nanopowders are available as powders and dispersions. These nanoparticles exhibit antibacterial, anti-corrosive, antifungal and UV filtering properties. Zinc is a Block D, Period 4 element while Oxygen is a Block P, Period 2 element. Some of the synonyms of zinc oxide nanoparticles are oxydatum, zinci oxicum, permanent white, ketozinc and oxozinc.

Chemical Properties

The chemical properties of zinc oxide nanoparticles are as per the table below.

Chemical Data	
Chemical symbol	ZnO
CAS No	1314-13-2
Group	Zinc 12Oxygen 16
Electronic configuration	Zinc [Ar] $3d^{10}$
$4s^2$Oxygen [He] $2s^2$	$2p^4$

Chemical Composition	
Element	Content (%)
Zinc	80.34
Oxygen	19.6

Physical Properties

The table given below highlights the physical properties of zinc oxide nanopowders.

Properties	Metric	Imperial
Density	5600 kg/m^3	0.202 lb/in^3
Molar Mass	81.40 g/mol	-

Thermal Properties

Given below are the thermal properties of zinc oxide nanoparticles.

Properties	Metric	Imperial
Melting point	1975°C	3587°F
Boiling point	2360°C	4280°F

Applications

Some of the applications of zinc oxide nanoparticles are given below.

- Zinc oxide is used in the manufacture of rubber and cigarettes (used as a filter).
- Popularly known calamine lotion is made out of zinc oxide powder. It is also used in a host of other creams and ointments that are used to treat skin diseases.
- As an additive in the manufacture of concrete.
- Ceramic industry has a number of uses for zinc oxide powder.
- It is also used as an additive in food products such as breakfast cereals.
- Various paints use zinc oxide as a coating agent.

Unimolecular Rectifier

A Unimolecular rectifier is a single organic molecule which functions as a rectifier (one-way conductor) of electric current. The idea was first proposed in 1974 by Arieh (later Ari) Aviram, then at IBM, and Mark Ratner, then at New York University. Their publication was the first serious and concrete theoretical proposal in the new field of molecular electronics (UE).

Their proposed rectifying molecule was designed so that electrical conduction within it would be favoured from the electron-rich subunit or moiety (electron donor) to an electron-poor moiety (electron acceptor), but disfavoured (by several electron volts) in the reverse direction.

Research

Many potential rectifying molecules were studied by the groups of Robert Melville Metzger, Charles A. Panetta, and Daniell L. Mattern (University of Mississippi) between 1981 and 1991, but were not tested successfully for conductivity.

This proposal was verified in two papers in 1990 and 1993 by the groups of John Roy Sambles (University of Exeter, UK) and Geoffrey Joseph Ashwell (Cranfield University now at the Lancaster University, UK), using a monolayer of hexadecylquinolinium tricyanoquinodimethanide sandwiched between dissimilar metal electrodes (magnesium and platinum) and then confirmed in three papers in 1997 and 2001 by Metzger (now at the University of Alabama) and coworkers, who used identical metals (first aluminium, then gold).

These papers use Langmuir-Blodgett monolayers (one molecule thick) with an estimated 10^{14} to 10^{15} molecules measured in parallel. About nine similar rectifiers of vastly different structure have been found by Metzger's group between 1997 and 2006. Some more perylene based organic rectifiers with PEG (polyethylene glycol) swallowtails have been synthesized in Mattern's lab by Ramakrishna Samudrala. These rectifiers would allow the rectification to be measured with flexibility.

Single molecules bonded covalently to gold have been studied by scanning tunneling spectroscopy and some of them are unimolecular rectifiers, studied as single molecules, as shown by the groups of Luping Yu (University of Chicago) and Ashwell (later at Lancaster University, UK).

Aims

The driving idea in UE (also called molecular-scale electronics) is that properly designed "electroactive" molecules, of between 1 and 3 nm in length, can supplant silicon-based devices to reduce circuit component sizes, providing concomitant increase in maximum integrated circuit speeds. However, amplification had not been realised as of 2012, and the chemical interactions between metal electrodes and molecules are complex.

Other Approaches

Nanoionics studies the transport of ions rather than electrons in nanoscale systems.

Nanophotonics studies the behaviour of light on the nanoscale, and has the goal of developing devices that take advantage of this behaviour.

Nanoelectronic Devices

Current high-technology production processes are based on traditional top down strategies, where nanotechnology has already been introduced silently. The critical length scale of integrated circuits is already at the nanoscale (50 nm and below) regarding the gate length of transistors in CPUs or DRAM devices.

Computers

Nanoelectronics holds the promise of making computer processors more powerful than are possible with conventional semiconductor fabrication techniques. A number of approaches are currently being researched, including new forms of nanolithography, as well as the use of nanomaterials such as nanowires or small molecules in place of traditional CMOS components. Field effect transistors have been made using both semiconducting carbon nanotubes and with heterostructured semiconductor nanowires.

In 1999, the CMOS transistor developed at the Laboratory for Electronics and Information Technology in Grenoble, France, tested the limits of the principles of the MOSFET transistor with a diameter of 18 nm (approximately 70 atoms placed side by side). This was almost one tenth the size of the smallest industrial transistor in 2003 (130 nm in 2003, 90 nm in 2004, 65 nm in 2005 and 45 nm in 2007). It enabled the theoretical integration of seven billion junctions on a €1 coin. However, the CMOS transistor, which was created in 1999, was not a simple research experiment to study how CMOS technology functions, but rather a demonstration of how this technology functions now that we ourselves are getting ever closer to working on a molecular scale. Today it would be impossible to master the coordinated assembly of a large number of these transistors on a circuit and it would also be impossible to create this on an industrial level.

Memory Storage

Electronic memory designs in the past have largely relied on the formation of transistors. However, research into crossbar switch based electronics have offered an alternative using reconfigurable interconnections between vertical and horizontal wiring arrays to create ultra high density memories. Two leaders in this area are Nantero which has developed a carbon nanotube based crossbar memory called Nano-RAM and Hewlett-Packard which has proposed the use of memristor material as a future replacement of Flash memory.

An example of such novel devices is based on spintronics. The dependence of the resistance of a material (due to the spin of the electrons) on an external field is called magnetoresistance. This effect can be significantly amplified (GMR - Giant Magneto-Resistance) for nanosized objects, for example when two ferromagnetic layers are separated by a nonmagnetic layer, which is several nanometres thick (e.g. Co-Cu-Co). The GMR effect has led to a strong increase in the data storage density of hard disks and made the gigabyte range possible. The so-called tunneling magnetoresistance (TMR) is very similar to GMR and based on the spin dependent tunneling of electrons through adjacent ferromagnetic layers. Both GMR and TMR effects can be used to create a non-volatile main memory for computers, such as the so-called magnetic random access memory or MRAM.

Novel Optoelectronic Devices

In the modern communication technology traditional analog electrical devices are increasingly replaced by optical or optoelectronic devices due to their enormous bandwidth and capacity, respectively. Two promising examples are photonic crystals and quantum dots. Photonic crystals are materials with a periodic variation in the refractive index with a lattice constant that is half the wavelength of the light used. They offer a selectable band gap for the propagation of a certain wavelength, thus they resemble a semiconductor, but for light or photons instead of electrons. Quantum dots are nanoscaled objects, which can be used, among many other things, for the construction of lasers. The advantage of a quantum dot laser over the traditional semiconductor laser is that their emitted wavelength depends on the diameter of the dot. Quantum dot lasers are cheaper and offer a higher beam quality than conventional laser diodes.

Displays

The production of displays with low energy consumption might be accomplished using carbon nanotubes (CNT). Carbon nanotubes are electrically conductive and due to their small diameter of several nanometres, they can be used as field emitters with extremely high efficiency for field emission displays (FED). The principle of operation resembles that of the cathode ray tube, but on a much smaller length scale.

Quantum Computers

Entirely new approaches for computing exploit the laws of quantum mechanics for novel quantum computers, which enable the use of fast quantum algorithms. The Quantum computer has quantum bit memory space termed "Qubit" for several computations at the same time. This facility may improve the performance of the older systems.

Radios

Nanoradios have been developed structured around carbon nanotubes.

Energy Production

Research is ongoing to use nanowires and other nanostructured materials with the hope to create cheaper and more efficient solar cells than are possible with conventional planar silicon solar cells. It is believed that the invention of more efficient solar energy would have a great effect on satisfying global energy needs. There is also research into energy production for devices that would operate *in vivo*, called bio-nano generators. A bio-nano generator is a nanoscale electrochemical device, like a fuel cell or galvanic cell, but drawing power from blood glucose in a living body, much the same as how the body generates energy from food. To achieve the effect, an enzyme is used that is capable of stripping glucose of its electrons, freeing them for use in electrical devices. The average person's body could, theoretically, generate 100 watts of electricity (about 2000 food calories per day) using a bio-nano generator. However, this estimate is only true if all food was converted to electricity, and the human body needs some energy consistently, so possible power generated is likely much lower. The electricity generated by such a device could power devices embedded in the body (such as pacemakers), or sugar-fed nanorobots. Much of the research done on bio-nano generators is still experimental, with Panasonic's Nanotechnology Research Laboratory among those at the forefront.

Medical Diagnostics

There is great interest in constructing nanoelectronic devices that could detect the concentrations of biomolecules in real time for use as medical diagnostics, thus falling into the category of nanomedicine. A parallel line of research seeks to create nanoelectronic devices which could interact with single cells for use in basic biological research. These devices are called nanosensors. Such miniaturization on nanoelectronics towards in vivo proteomic sensing should enable new approaches for health monitoring, surveillance, and defense technology.

Bibliography

Altmann, J. & Gubrud, M.A.: *Discovering the Nanoscale*, IOS Press, Amsterdam, 2004.

Andrew, N.: *Foundations of Nanomechanics: from Solid-state Theory to Device Applications*, Berlin, New York, 2003.

Attebery, B.: *Nanoculture: Implications of the New Technoscience*, Intellect Books, Bristol, 2004.

Bainbridge, W.S. : *Societal Implications of Nanoscience and Nanotechnology*, Dordrecht, Kluwer, 2001.

Bast, R.C. Jr.: *Cancer Medicine*, Decker, Hamilton, 2000.

Berube, D.M.: *Discovering the Nanoscale*, IOS Press, Amsterdam, 2004.

Brosnan, J.: *The Primal Scream: A History of Science Fiction Film*, Orbit Books, London, 1991.

Carroll, J.S.: *Social Science Research Methods for Assessing Societal Implications of Nanotechnology*, Dordrecht, Kluwer, 2001.

Caviedes, C.: *El Nino in History: Storming Through the Ages*, University Press of Florida, Florida, 2001.

Credi, A.: *Molecular Devices and Machines: a Journey into the Nano World*, Wiley-VCH, Weinheim, 2003.

Crow, M. & Sarewitz, D.: *Societal Implications of Nanoscience and Nanotechnology*, Dordrecht: Kluwer, 2001.

Daniel Ratner: *Nanotechnology: a Gentle Introduction to the Next Big Idea*, Upper Saddle River, Prentice Hall, 2003.

Dobzhanshy, T.: *Genetics of the Evolutionary Process*, Columbia University Press, New York, 1970.

Drexler, K. Eric: *Engines of Creation: The Coming Era of Nanotechnology*, Anchor Books, New York, 1986.

Dyson, Tim: *India's Historical Demography: Studies in Famine, Disease, and Society*, Riverdale, Westwood, MA, 1989.

Eamonn Gearon: *The Sahara: A Cultural History*, Signal Books, UK, 2011.

Frank J. Owens: *Introduction to Nanotechnology*, Hoboken, J. Wiley, 2003.

Glimell, H.: *Dynamics of the Emerging Field of Nanoscience*, Dordrecht, Kluwer, 2001.

Glimell, H.: *Societal Implications of Nanoscience and Nanotechnology*, Kluwer, Dordrecht, 2001.

Haber, L. F.: *The Chemical Industry: 1900-1930, International Growth and Technological Change*, Oxford: Claredon Press, 1971.

Hessenbruch, A.: *Nanotechnology and the Negotiation of Novelty*, Amsterdam, IOS Press, 2004.

James Lewis: *Prospects in Nanotechnology: Toward Molecular Manufacturing,* John Wiley & Sons, New York, 1995.

Karn, Barbara: *Nanotechnology and the Environment: Applications and Implications,* American Chemical Society, Washington D.C., 2004.

Keblinski, P. : *Encyclopedia of Nanoscience and Nanotechnology*, American Scientific, Los Angeles, Calif, USA, 2004.

Liu W.K., Karpov E.G., Park H.S.: *Nano Mechanics and Materials,* Wiley, Weinheim, 2006.

Mayr, E.: *Change of Genetic Environment and Evolution*, Allen and Unwin, London, 1954.

Mody, C.C.M.: *How Probe Microscopists Became Nanotechnologists, Discovering the Nanoscale*, IOS Press, Amsterdam, 2004.

Norio Taniguchi: *Nanotechnology: Integrated Processing Systems for Ultra-precision and Ultra-fine Products,* Oxford University Press, New York, 1996.

Orth, R. N.: *Nanometer Scale Antibody Patterning for Directed cell Immobilization and Stimulation,* Cornell University, Ithaca, 2002.

Pressman, J.: *Nano Narrative: A Parable from Electronic Literature*, Intellect Books, Bristol, 2004.

Pressman, J.: *Nanoculture: Implications of the New Technoscience*, Intellect Books, Bristol, UK, 2004.

Ratner, Daniel: *Nanotechnology: a Gentle Introduction to the Next Big Idea*, Prentice Hall, Upper Saddle River, 2003.

Robison, W.L.: *Discovering the Nanoscale*, IOS Press, Amsterdam, 2004.

Sarewitz, D.: *Nanotechnology and Societal Transformation*, Dordrecht, Kluwer, 2001.

Stephen M. Goodnick: *Transport in Nanostructures,* Cambridge University Press, Cambridge, 1997.

Tenner, E.: *Nanotechnology and Unintended Consequences*, , Dordrecht, Kluwer, 2001.

Vasudev Bhatnagar: *Cell Science and Technology*, Campus Books, Delhi, 2009.

Vermaas, P.E.: *Discovering the Nanoscale*, IOS Press, Amsterdam, 2004.

Weil, V.: *Ethical Issues in Nanotechnology*, Dordrecht, Kluwer, 2001.

Index

L

M

N

O

P

R

S

T

U

V

W

Z

❑❑❑